T0269247

CAMBRIDGE LIBRARY COLLECTION

Books of enduring scholarly value

Physical Sciences

From ancient times, humans have tried to understand the workings of the world around them. The roots of modern physical science go back to the very earliest mechanical devices such as levers and rollers, the mixing of paints and dyes, and the importance of the heavenly bodies in early religious observance and navigation. The physical sciences as we know them today began to emerge as independent academic subjects during the early modern period, in the work of Newton and other 'natural philosophers', and numerous sub-disciplines developed during the centuries that followed. This part of the Cambridge Library Collection is devoted to landmark publications in this area which will be of interest to historians of science concerned with individual scientists, particular discoveries, and advances in scientific method, or with the establishment and development of scientific institutions around the world.

Weather Lore

Richard Inwards (1840–1937) trained as a mining engineer, working on projects in Europe and South America (his book on Tiwanaku in Bolivia, *The Temple of the Andes*, is also reissued in the Cambridge Library Collection). A fellow of the Royal Meteorological Society and the Royal Astronomical Society, Inwards became well known in scientific circles. *Weather Lore* was first published in 1869, with this 1893 second edition including new entries from the United States. Compiled from sources as diverse as Hesiod, the Bible and Francis Bacon, the collection includes the notable observations that 'if spaniels sleep more than usual, it foretells wet weather', but 'if rats are more restless than usual, rain is at hand'. Often entertaining, always fascinating, the book does not pretend to be scientifically accurate; as the author was to remark later, 'no human being can correctly predict the weather, even for a week to come'.

Cambridge University Press has long been a pioneer in the reissuing of out-of-print titles from its own backlist, producing digital reprints of books that are still sought after by scholars and students but could not be reprinted economically using traditional technology. The Cambridge Library Collection extends this activity to a wider range of books which are still of importance to researchers and professionals, either for the source material they contain, or as landmarks in the history of their academic discipline.

Drawing from the world-renowned collections in the Cambridge University Library and other partner libraries, and guided by the advice of experts in each subject area, Cambridge University Press is using state-of-the-art scanning machines in its own Printing House to capture the content of each book selected for inclusion. The files are processed to give a consistently clear, crisp image, and the books finished to the high quality standard for which the Press is recognised around the world. The latest print-on-demand technology ensures that the books will remain available indefinitely, and that orders for single or multiple copies can quickly be supplied.

The Cambridge Library Collection brings back to life books of enduring scholarly value (including out-of-copyright works originally issued by other publishers) across a wide range of disciplines in the humanities and social sciences and in science and technology.

Weather Lore

A Collection of Proverbs, Sayings, and Rules Concerning the Weather

RICHARD INWARDS

CAMBRIDGE
UNIVERSITY PRESS

University Printing House, Cambridge, CB2 8BS, United Kingdom

Cambridge University Press is part of the University of Cambridge.
It furthers the University's mission by disseminating knowledge in the pursuit of education, learning and research at the highest international levels of excellence.

www.cambridge.org
Information on this title: www.cambridge.org/9781108077620

This edition first published 1893
This digitally printed version 2015

ISBN 978-1-108-07762-0 Paperback

The original edition of this book contains a number of oversize plates which it has not been possible to reproduce to scale in this edition. They can be found online at www.cambridge.org/9781108077620

WEATHER LORE

ARISE : PRAY : WORK :

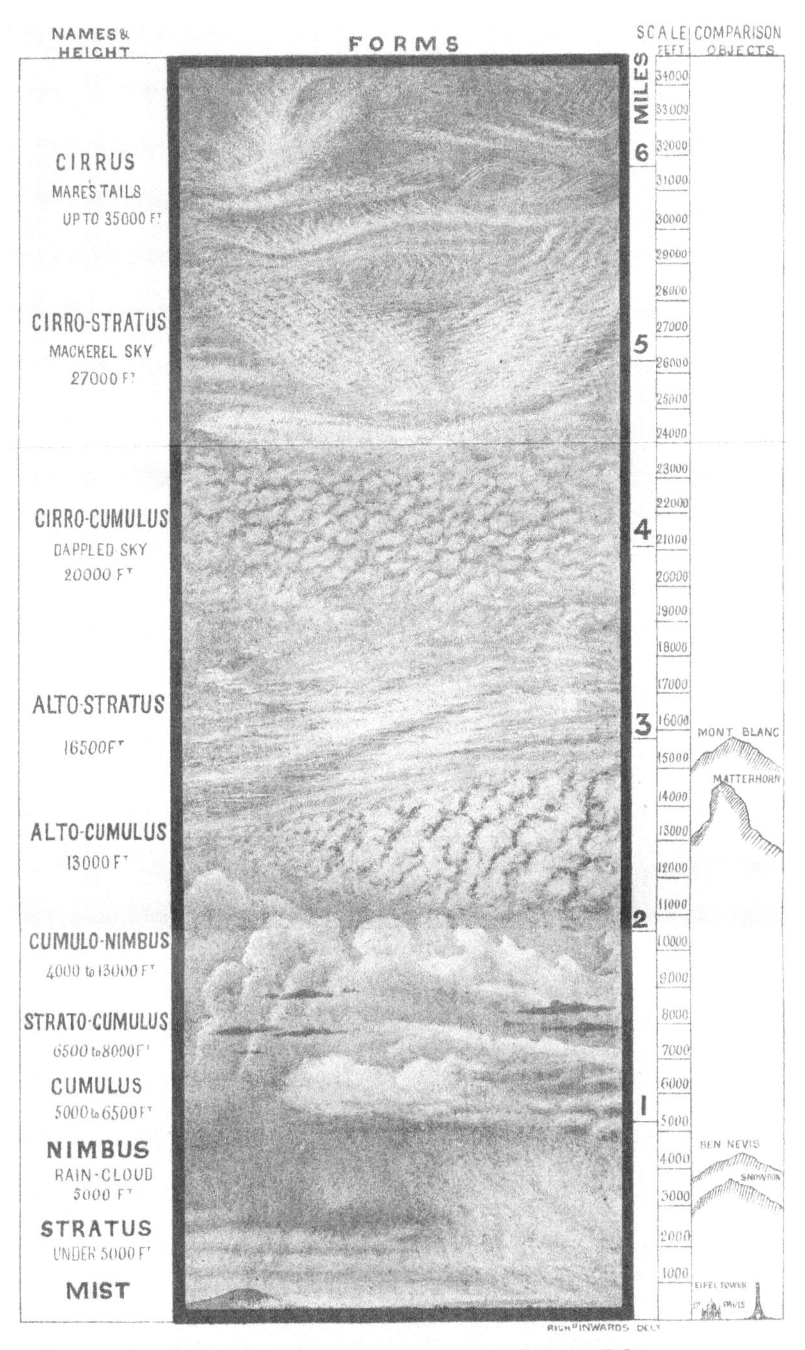

FORMS, HEIGHTS & NAMES OF CLOUDS.

Compiled by permission from Dr Karl Singer's Photographs (Wolkentafeln Munich 1892)

The material originally positioned here is too large for reproduction in this reissue. A PDF can be downloaded from the web address given on page iv of this book, by clicking on 'Resources Available'.

The material originally positioned here is too large for reproduction in this reissue. A PDF can be downloaded from the web address given on page iv of this book, by clicking on 'Resources Available'.

WEATHER LORE

A COLLECTION OF

Proverbs, Sayings, and Rules

CONCERNING THE WEATHER

COMPILED AND ARRANGED BY

RICHARD INWARDS, F.R.A.S.

FELLOW OF THE ROYAL METEOROLOGICAL SOCIETY; AUTHOR OF "THE TEMPLE OF THE ANDES"

LONDON

ELLIOT STOCK, 62, PATERNOSTER ROW

1893

INTRODUCTION.

The state of the weather is almost the first subject about which people talk when they meet, and it is not surprising that a matter of such importance to comfort, health, prosperity, and even life itself, should form the usual text and starting-point for the conversation of daily life.

From the earliest times, hunters, shepherds, sailors, and tillers of the earth have from sheer necessity been led to study the teachings of the winds, the waves, the clouds, and a hundred other objects from which the signs of coming changes in the state of the air might be foretold. The weather-wise amongst these primitive people would be naturally the most prosperous, and others would soon acquire the coveted foresight by a closer observance of the same objects from which their successful rivals guessed the proper time to provide against a storm, or reckoned on the prospects of the coming crops. The result has been the framing of a rough set of rules, and the laying down of many "wise saws," about the weather, and the freaks to which it is liable. Some of these observations have settled down into the form of proverbs; others have taken the shape of rhymes; while many are yet floating about, unclaimed and unregistered, but passed from mouth to mouth, as mere records of facts, varying in verbal form according to local idioms, but owning a common origin and purport.

Many weather proverbs contain evidence of keen observation and just reasoning, but a great number are the offspring of the common tendency to form conclusions from a too limited observation of facts. Even those which have not been confirmed by later experience will be interesting, if only to show the errors into which men may be led by seeing nature with eyes half closed by prejudice or superstition. It has seemed to me desirable that all this "fossil wisdom" should be collected, and I have endeavoured in this book to present in a systematic form all the current weather lore applicable to the climate of the British Isles.

This work is not intended to touch the philosophical aspect of the subject, but it is hoped that its perusal may lead some people to study the weather, not by mere "rule of thumb," as their fathers did, but by intelligent observation, aided by all the niceties of the scientific means now fortunately at the command of every one.

This collection comprises only those proverbs, sayings, or rules in some way descriptive or prophetic of the weather and its changes, and does not include those in which the winds, sun, and clouds are only brought in for purposes of comparison and illustration—such, for instance, as "Always provide against a *rainy* day," "Every *cloud* has a silver lining," and others in which the weather is only incidentally or poetically mentioned. Some rhymes have been rejected on account of their being manifestly absurd or superstitious, but the reader will see that much latitude has been allowed in this respect, and, as a rule, all those which may possibly be true will be found in these pages. Predictions as to the peace of the realm, the life and death of kings, etc., founded on the state of the weather for particular days, have of course been left out, as unworthy of remembrance.

A few of the rules here presented will very possibly be found to contradict each other, but the reader will judge between them, and assign each its proper value. With regard to those from foreign sources, I have only been able to give a few which seem in some measure applicable to our climate, and it will be seen that even these have lost a great deal of their point in the process of translation. A great many proverbs about the weather come from Scotland, very few from Ireland.

I have registered the various extracts in the order which seemed most convenient for reference, generally giving precedence to the subjects on which they were the most numerous. Respecting the sources from which they have been derived, I have, of course, availed myself of the collections of general proverbs by Kelly, Howell, Henderson, and Ray. The collection by the latter author, which is usually considered the most complete, only contains, however, eighty-seven adages, which have been transcribed into this volume as weather proverbs proper. A much greater number have appeared in the estimable *Notes and Queries*, under the head of "Folk Lore," and a few have been gleaned from Hone's *Every-Day Book* and other volumes of a similar class. The rest have, for tee most part, come under my personal notice, or have been communicated by esteemed correspondents, who are now heartily thanked. A full list of the various authors to whom I am indebted will be found in the appendix.

The Bible has handed down to us many proofs of the repute in which weather wisdom was held by the ancients, and it is clear that some of the sacred writers were keen observers of the signs of the sky. The writings of Job are rich in this respect, and contain many allusions to the winds, clouds, and tempests. The New Testament also records some sound weather law, and in one instance Christ Himself has not thought it unworthy of Him to confirm a popular adage about a cloud rising in the west and foreshowing rain; for after mentioning the saying, He has added, "And so it is." The texts referring to the weather have therefore been inserted where appropriate. In their proper places, too, will be found quotations from learned authors, amongst whom Shakespeare holds a prominent place. The admirers of that poet "for all time" will not be surprised to find that he has said, in his own way, nearly all that was known on the subject of the "skyey influences" in the age in which he lived. Virgil, Bacon, Thompson, and other less famous men will be shown to have contributed something to the common stock of information on this subject. Some sound Saxon weather lore comes also from the mouth of the Shepherd of Banbury, who, in the last century, wrote a short list of outdoor signs of coming changes in the state of the air.

The collection of Scottish weather proverbs by Sir A. Mitchell has furnished me with a few of the shrewdest adages from that country, and the list published by Mr. M. A. Denham for the Percy Society has yielded some not met with in any other place.

In this Second Edition I have been able, by courteous permission of Brigadier-General Greely, of the Washington Signal Office, to incorporate a great number of American and other proverbs, which have been collected for the United States Signal Service by Major Dunwoody.

Mr. P. Dudgeon, of Cargen, has been kind enough to make many important corrections to the Scottish sayings which appear in this work.

I desire also to acknowledge my great obligation to the Rev. C. W. Empson for many kind hints and corrections, and to thank Mr. G. J. Symons for having kindly allowed me the use of his priceless meteorological library.

As it has been impossible to collect all the local weather proverbs current in different parts of the country, I shall feel obliged to any courteous reader who will communicate such as have been omitted, so that a future edition of this work may be rendered more complete in this respect. It would be strange if all the observations here brought for the first time to a common focus did not cast a new ray or two of light on the point to which they have all been directed. Out of so many shots some must hit the mark, though the reader must be warned that even in this "multitude of counsel" there is not absolute safety. These predictions are after all but gropings in the dark; and although skilled observers, armed with the delicate instruments contrived by modern science, may be able to forecast with some success the weather for a few hours, yet with respect to the coming months and seasons, or the future harvests and vintages, the learned meteorologist is only on a level with the peasant who watches from the hilltop the "spreadings and driftings of the clouds," or hazards his rude weather guesses from the behaviour of his cattle or the blossoming of the hedge flowers which he daily sees.

It is, perhaps, worth mentioning, with respect to those proverbs concerning the weather of particular days, that, on account of the re-formation of the calendar, a great many of these sayings must be held to refer to times a little later than the dates now affixed. Notwithstanding this, I have retained the dates which I find by custom attached to the adages, as it is now impossible to say how long before the alteration of the calendar they took their rise. Of course the real discrepancy will depend on the date of origin, as, in the case of any proverb having been current in the time of Julius Cæsar, its date would refer to the same part of the earth's orbit as at present, while the "Saints' Day" proverbs which have been concocted in the Middle Ages would require a correction depending upon the error of the calendar which had accumulated at their date of origin. This alone would account for the uncertain value of all this class of predictions. The list of times for the flowering of plants must also be taken with many allowances, on account of the varying soil and climate of the different parts of the kingdom from which the information was collected.

Should the reader ask, as he naturally may, to what practical result does all this tend, and how from it he may venture to predict the coming weather, I can only recommend him to try and imbibe the general spirit of the rules and adages, to watch the clouds from a high place, to examine the published weather diagrams, and by collating them try to find where similar results have followed similar indications, and by all the instrumental means he can, go on measuring and gauging heat, pressure, rain, wind, and moisture, in the hope that he may some day arrive at the semblance of a definite law, and the certainty that he is pursuing an interesting and ever-improving study.

As for this book, it aims at no more than being a manual of outdoor weather wisdom seen from its traditional and popular side, without pretending to any scientific accuracy. Meteorology itself, especially as regards English weather, is very far from having reached the phase of an exact science.

RICHARD INWARDS.

BARTHOLOMEW VILLAS,
LONDON, N.W.

CONTENTS.

[*For full Index, see page* 174]

Weather in General.

The weather rules the field.—SPANISH. *Weather.*

'Tis not the husbandman, but the good weather, that makes the corn grow.—T. FULLER. *Good weather.*

In the reign of Henry VIII. a proclamation was made against the almanacks which transmitted the belief in saints ruling the weather. *Proclamation against weather saints.*

Better it is to rise betimes
And make hay while the sun shines,
Than to believe in tales and lies
Which idle people do devise. *Sunshine.*

Of Albion's glorious Ile, the wonders whilst I write,
The sundry varying soyles, the pleasures infinite;
Where heat kills not the cold, nor cold expells the heat,
Ne calmes too mildly small, nor winds too roughly great;
Nor night doth hinder day, nor day the night doth wrong,
The summer not too short, the winter not too long.
DRAYTON. *English climate.*

Husbandry depended on the periodical rains; and forecasts of the weather, with a view to make adequate provision against a coming deficiency, formed a special duty of the Bráhmans. The philosopher who erred in his predictions observed silence for the rest of his life.
W. W. HUNTER. *Weather prophecy (Indian).*

There are many weathers in five days, and more in a month.
NORWAY. *Weathers.*

Those that are weather wise
Are rarely otherwise.—CORNWALL. *Weather prophets.*

Whether the weather be fine or wet,
Always water when you set. *Sowing weather.*

Weather, wind, women, and fortune change like the moon.
FRENCH. *Weather changes.*

Weather changes.

When an opinion once obtains that a change of the weather happens at certain times, the change is expected, and as often as it takes place the remembrance of it remains; but we soon forget the number of times it fails.—JOHN MILLS, F.R.S. (ESSAY ON THE WEATHER).

Weather fine.

If the weather is fine, put on your cloak.
If it is wet, do as you please.—FRENCH.

Weather signs.

Aratus says: "Do not neglect any of these [weather] signs, for it is good to compare a sign with another sign: if two agree, have hope, but be assured still more by a third."
PRINCE.

Weather rhyme.

"Well, Duncombe, how will be the weather?"
"Sir, it looks cloudy altogether;
And coming across our Houghton Green,
I stopped and talked with old Frank Beane
While we stood there, sir, old Jan Swain
Went by, and said he knowed 'twould rain;
The next that came was Master Hunt,
And he declared he knew it wouldn't;
And then I met with Farmer Blow—
He plainly said he didn't know.
So, sir, when doctors disagree,
Who's to decide it—you or me?"

[This is a village rhyme written in the last century, and well known in Bedfordshire, where all the names are still found.]

Weather bad.

Shepherd.—"Weel, do ye ken, sir, that I never saw in a' my born days what I could wi' a safe conscience hae ca'd bad weather? The warst has aye some redeemin' quality about it that enabled me to thole it without yaumerin [murmuring]. Though we may na be able to see, we can aye think of the clear blue lift. Weather, sir, aiblins no to speak very scientially in the way o' meteorological observation—but rather in a poetical, that is, a religious spirit—may be defined, I jalouse [suspect], 'the expression o' the fluctuations and modifications of feeling in the heart o' the heevens made audible and visible and tangible on their face and bosom.' That's weather."—PROFESSOR WILSON.

Weather and health.

The common feelings of every man will convince him, if he will attend to them, of the superior advantages health derives from a pure and temperate atmosphere; for while troubled, tempestuous, foul, rough, and impetuous weather prevails, while the days are cloudy and the nights damp, the mind becomes tetrick [perverse], sad, peevish, angry, dull, and melancholy; but while the western gales blow calmly over our heads, and the sun shines mildly from the skies, all nature looks alert and cheerful.

Weather and health.

> Thus when the changeful temper of the skies
> The rare condenses, the dense rarefies,
> New motions on the altered air impress't,
> New images and passions fill the breast;
> Then the glad birds in tender concert join,
> Then croaks the exulting rook, and sport the lusty kine.
>
> *Virgil's "Georgics," Book I., Line* 490.

Weather works on all in different degrees, but most on those who are disposed to melancholy. The devil himself seems to take the opportunity of foul and tempestuous weather to agitate our spirits and vex our souls; for as the sea waves, so are the spirits and humours in our bodies tossed with tempestuous winds and storms.—BURTON'S "ANATOMY OF MELANCHOLY," CHAP. III.

Weather prayers.

In one of Lucian's *Dialogues* there is an account of a couple of countrymen,—one pouring into the right ear of the god a petition that not a drop of rain may fall before he has completed his harvest; while another peasant, equally importunate, whispers into the left ear a prayer for immediate rain, in order to bring on a backward crop of cabbages.

Weather madness.

The astronomer, in Dr. Johnson's *Rasselas*, goes mad on the subject of the weather, which he fully believes he can control; and there have not been wanting in modern times sages who believed themselves equally potent, and some of them have gone the length of offering to predict the weather for any future time on payment of a fee, whilst the moderate price of sixpence was indicated as necessary for a single day's prophecy.

Times and Seasons.

Amongst the first attempts at weather guesses, those concerning the seasons and their probable fitness for agriculture, the breeding of animals, or the navigation of the seas would take a prominent place. The weather during the winter and spring seems to have been narrowly watched, and the chances of a good harvest, a fat pasture, or a loaded orchard inferred from the experience of previous years, combined with a fair reliance upon fortune. Some of these predictions, though not strengthened by modern observation, are not to be altogether despised or thrown aside. They at least show us what kind of weather our forefathers wished to take place and thought most useful at the times to which they refer. The sayings of French, Scotch, and English agree in many particulars—such, for instance, as those referring to Candlemas Day and the early part of February generally. It seems that, according to the notions of our ancestors, this part of the year could not be too cold, and no statistical evidence will ever make our farmers believe that a warm Christmas bodes well for an English harvest, or that a dry year ever did harm to the conntry. Some of these old sayings are also interesting as perhaps indicating the slowly changing climate of this country, and it is not unlikely that at some distant date most of the predictions will be found inapplicable. Particular saints' days have also been selected as exerting special influence over the weather, and here we are constantly treading on the fringes of the veil of superstition, spread by ignorance over all matters about which but little certain knowledge existed. There are, however, still believers in St. Swithin and St. Valentine as weather prophets; and if their favourites do sometimes fail to bring the expected changes, they have at least no worse guides than those furnished by the Old Moore's and Zadkiel's of modern times.

It has been thought advisable to admit the proverbs concerning the proper seasons for sowing, etc.; and a table of the times of the flowering of certain well-known plants has been added, so that the progress of the seasons may be watched by observing the punctuality of the vegetable world in heralding their approach.

NOTE ON NEW STYLE.—*In considering the weather proverbs regarding certain days, it must be remembered that the New Style was first adopted September* 2*nd,* 1752, *eleven days being retrenched from the calendar:* i.e., *August* 22*nd,* 23*rd,* 24*th,* 25*th,* 26*th,* 27*th,* 28*th,* 29*th,* 30*th,* 31*st, and September* 1*st,* 1752, *had no existence in England.*

YEAR.

Good. A good year is always welcome.—ICELAND.

Year. Do not abuse the year till it has passed.—SPAIN.

Old year. If the old year goes out like a lion, the new year will come in like a lamb.

Harvest. The harvest depends more on the year than on the field.
DENMARK.

Dry. A dry year never beggars the master.—FRENCH.

A dry year never starves itself.

Whoso hath but a mouth
Will ne'er in England suffer drought.

If there be neither snow nor rain,
Then will be dear all kinds of grain.

Wet. A bad year comes in swimming.—FRENCH.

After a wet year a cold one.

Rainy year,
Fruit dear.—HAUTE LOIRE.

Wet and dry. Wet and dry years come in triads.

[Year.]

Misty year, year of cornstalks.—SPANISH. *Misty.*

Year of frosts, year of cornstacks.—SPANISH. *Frosty.*

Frost year,
Fruit year.—EURE ET LOIRE.

Frost year, wheat year.—FRANCE.

Year of snow,
Fruit will grow.—MILAN. *Snowy.*

A snow year, a rich year.

Snow year, good year.

A year of snow, a year of plenty.—SPANISH AND FRENCH.

A year of wind is good for fruit.—CALVADOS. *Windy.*

Acorn year, purse year.
Fig year, worse year.—SPANISH. *Acorns and figs.*

A good nut year, a good corn year. *Nuts.*

Year of nuts,
Year of famine.—FRANCE (HAUTE MARNE).

A good hay year, a bad fog year. *Hay.*

A year of grass good for nothing else.—SWITZERLAND. *Grass.*

A pear year,
A dear year. *Pears.*

A cherry year,
A merry year.
A plum year,
A dumb year.—KENT. *Cherries and plums.*

In the year when plums flourish all else fails.—DEVONSHIRE. *Plums.*

Year of gooseberries, year of bottles [good vintage].—FRANCE. *Gooseberries.*

A haw year,
A braw year.—IRELAND AND SCOTLAND. *Haws.*

A haw year,
A snaw year.—SCOTLAND.

Year of mushrooms,
Year of poverty.—FRANCE (HAUTES PYRENEES). *Mushrooms.*

Year of radishes,
Year of health.—ARDÈCHE. *Radishes.*

Year of cockchafers, year of apples.—FRANCE. *Cockchafers.*

A cow year, a sad year;
A bull year, a glad year.—DUTCH. *Cows.*

Corn and horn go together. *Corn and cattle.*

Leap year was ne'er a good sheep year.—SCOTLAND. *Leap.*

SEASONS. A serene autumn denotes a windy winter; a windy winter, a rainy spring; a rainy spring, a serene summer; a serene summer, a windy autumn, so that the air on a balance is seldom debtor to itself.—LORD BACON.

Satire on seasons.

Spring. Slippy, drippy, nippy.
Summer. Showery, flowery, bowery.
Autumn. Hoppy, croppy, poppy.
Winter. Wheezy, sneezy, breezy.
ATTRIBUTED TO SYDNEY SMITH.

[Composed as a satirical mistranslation of the names given to the months at the time of the French Revolution.—G. F. CHAMBERS.]

Extreme. Extreme seasons are said to occur from the sixth to the tenth year of each decade, especially in alternate decades.

The first three days of any season rule the weather of that season.

The general character of the weather during the last twenty days of March, June, September, or December will rule the following season.

[SPRING.] Spring is both father and mother to us.—GALICIA.

Late.

A late spring
Is a great bless-ing.

A late spring never deceives.

Better late spring and bear, than early blossom and blast.

When the cuckoo comes to the bare thorn,
Sell your cow and buy your corn;
But when she comes to the full bit,
Sell your corn and buy your sheep.

i.e., A late spring is bad for cattle, and
An early spring is bad for corn.

Cold. If the spring is cold and wet, then the autumn will be hot and dry.

Dry. A dry spring, rainy summer.—FRANCE.

Damp. A wet spring, a dry harvest.

Spring rain damps, autumn rain soaks.—RUSSIA.

In spring a tub of rain makes a spoonful of mud.
In autumn a spoonful of rain makes a tub of mud.

The spring is not always green.

Day. An unseasonably fine day in spring or winter is called a pet day in Scotland. The fate of pets, they say, awaits it, and they look for spoilt weather on the morrow.

Seas. The spring openeth the seas for the sailors.—PLINY.

Thunder in spring
Cold will bring.

[Spring.] Thunder.

First thunder in spring,—if in the south, it indicates a wet season; if in the north, a dry season.

First thunder.

Early thunder, early spring.

Early thunder.

Lightning in spring indicates a good fruit year.

Lightning.

As the days grow longer,
The storms grow stronger.

Storms.

If there's spring in winter, and winter in spring,
The year won't be good for anything.

Spring in winter.

There are a hundred days of easterly wind in the first half of the year.—WEST OF ENGLAND.

Spring and summer.

[SUMMER.] *Moist.*

Generally a moist and cool summer portends a hard winter.
BACON.

An English summer, two hot days and a thunderstorm.

Stormy.

A dry summer never made a dear peck.

Dry.

A dry summer never begs its bread.—SOMERSET.

Whoso hath but a mouth
Will ne'er in England suffer drought.

Drought never bred dearth in England.

When the sand doth feed the clay,*
England woe and well a day;
But when the clay doth feed the sand,†
Then 'tis well for Angle-land.

Dry and wet.

After a famine in the stall, [Bad hay crop.]
Comes a famine in the hall. [Bad corn crop.]

A famine in England begins in the horse manger.

A hot and dry summer and autumn, especially if the heat and drought extend far into September, portend an open beginning of winter, and cold to succeed towards the latter part of the winter and beginning of spring.—BACON.

Hot and dry, extending into autumn.

One swallow does not make a summer.

Swallows.

Midsummer rain
Spoils hay and grain.

Rain.

Midsummer rain
Spoils wine stock and grain.—PORTUGUESE.

* As in a wet summer. † As in a dry summer.

[Summer.] Rainy

There can never be too much rain before midsummer.
SWEDEN.

Happy are the fields that receive summer rain.

If the summer be rainy, the following winter will be severe.

Fog.

In summer a fog from the south, warm weather; from the north, rain.

A summer fog is for fair weather.

Cool.

A cool summer and a light weight in the bushel.

Indian.

If we do not get our Indian summer in October or November, we shall get it in the winter.—UNITED STATES.

Summer and winter.

Summer comes with a bound; winter comes yawning.
FINLAND.

Days in summer.

As the days begin to shorten,
The heat begins to scorch them.

[AUTUMN.] Dry.

A fair and dry autumn brings in always a windy winter.
PLINY.

Dry vintage, good wine.—SPAIN.

Autumn and winter.

Clear autumn, windy winter;
Warm autumn, long winter.

Wet.

A wet fall indicates a cold and early winter.

Moist.

A moist autumn with a mild winter is followed by a cold and dry spring, retarding vegetation.

Fog.

Much fog in autumn, much snow in winter.

Thunder

Thunder in the fall indicates a mild, open winter.

Harvest short.

Short harvests make short addlings [earnings].—YORKSHIRE.

Long.

A long harvest, a little corn.

Fruits.

If you would fruit have,
You must bring the leaf to the grave.
[*i.e.*, transplant in autumn.]

Night.

The autumn night is changeable.—NORWAY.

[WINTER.] Dry.

Winter never rots in the sky.—ITALIAN.

Winter never died in a ditch.

Winter finds out what summer lays up.

Green.

A green winter makes a fat churchyard.

When there is a spring in the winter, or a winter in the spring, the year is never good.

Mild.

Summer in winter, and summer's flood,
Never boded an Englishman good.

An abundant wheat crop does not follow a mild winter.
FARMER, QUOTED IN "NOTES AND QUERIES,"
FEBRUARY 27TH, 1869.

A warm and open winter portends a hot and dry summer.
BACON.

[Winter.] Mild.

A warm winter and cool summer never brought a good harvest.—FRENCH.

Whae doffs his coat on winter's day
Will gladly put it on in May.—SCOTCH.

Early.

When winter begins early, it ends early.

An early winter,
A surly winter.

An early winter is surely winter.

An air' winter,
A sair winter.—SCOTLAND.

If the ice will bear a goose before Christmas, it will not bear a duck after.

Clear.

Neither give credit to a clear winter nor a cloudy spring.

Long.

Long winter and late spring are both good for hay and grain, but bad for corn and garden.

Rainy.

After a rainy winter follows a fruitful spring.

Floods.

Winter will not come till the swamps are full.
SOUTHERN UNITED STATES.

Fine day in.

An unusually fine day in winter is known locally as a "Borrowed Day," to be repaid with interest later in the season, known also as a "Weather Breeder," and by sailors as a "Fox."—ROPER.

Thunder.

Winter thunder,
A summer's wonder.

Winter thunder
Bodes summer's hunger.

Winter thunder and summer flood
Never boded an Englishman good.

Winter thunder,
Poor man's death, rich man's hunger.

Winter thunder,
Rich man's good and poor man's hunger.
[*i.e.*, it is good for fruit and bad for corn.]

Fog.

A winter fog
Will freeze a dog.

Frost.

Mony a frost and mony a thowe [thaw]
Soon maks mony a rotten yowe [ewe].

Snow.

Under water, dearth ;
Under snow, bread.

Dearth under water ;
Bread under snow.—ITALIAN.

Midwinter.

A seven-night before midwinter day and as much after, the sea is allayed and calm.—PLINY.

JANUARY.

Froze Janiveer,
Leader of the year;
Minced pies in van,
Calf's head in rear.—CHURCHILL.

The blackest month in all the year
Is the month of Janiveer.

A favourable January brings us a good year.

The month of January is like a gentleman (as he begins, so he goes on).—SPANISH.

Bright. In January if the sun appear,
March and April pay full dear.

Warm. January warm, the Lord have mercy!

A summerish January, a winterish spring.

Mild. If grain grows in January, there will be a year of great need.

Grass. If you see grass in January,
Lock your grain in your granary.

If the grass grow in Janiveer,
It grows the worse for it all the year.

Flowers. January flowers do not swell the granary.—SPANISH.

Blossoms. January blossoms fill no man's cellar.—PORTUGUESE.

Birds. If birds begin to whistle in January, frosts to come.—RUTLAND.

Gnats. When gnats swarm in January, the peasant becomes a beggar.—DUTCH.

Mild. If January calends be summerly gay,
It will be winterly weather till the calends of May.

Spring. A January spring is worth naething.—SCOTCH.

Dry. Dry January, plenty of wine.

Wet. A wet January, a wet spring.

Is January wet?—the barrel remains empty.

A wet January is not so good for corn, but not so bad for cattle.—SPANISH AND PORTUGUESE.

January wet, no wine you get.

Have rivers much water in January?—then the autumn will forsake them. But are they small in January?—then brings the autumn surely much wine.—SOUTH EUROPE.

In January much rain and little snow is bad for mountains, valleys, and trees.

Much rain in January, no blessing to the fruit.

Thaw. Always expect a thaw in January.

Fog. Fog in January brings a wet spring.

If there is no snow before January, there will be the more in March and April. [*January.*] *Snow.*

Janiveer freeze the pot by the fier. *Cold.*

As the day lengthens,
So the cold strengthens.

A kindly, good Janiveer
Freezeth the pot by the fire.—TUSSER.

Jack Frost in Janiveer
Nips the nose of the nascent year.

Hoar-frost and no snow is hurtful to fields, trees, and grain. *Frost.*

When oak trees bend with snow in January, good crops may be expected. *Oaks.*

If January could, he would be a summer month. *January.*
GREEK PROVERB, "THE CYCLADES,"
J. T. BENT, 1885, P. 86.

In January wane fell your timber.—SPANISH. *Timber.*

A January chicken is sold dearly or dies.—SPANISH. *Chickens.*

Thunder in January signifieth the same year great winds, plentiful of corn and cattle, peradventure.—BOOK OF KNOWLEDGE. *Thunder.*

January and February eat more than Madrid and Toledo.
SPANISH. *January and February.*

Generals January and February will fight for us.
CZAR NICHOLAS I.

January or February
Do fill or empty the granary.—FRENCH.

A cold January, a feverish February, a dusty March, a weeping April, and a windy May presage a good year and gay.
FRENCH. *January and following months.*

In January should sun appear,
March and April pay full dear. *January, March, and April.*

March in Janiveer,
Janiveer in March I fear. *January and March.*

Who in January sows oats
Gets gold and groats;
Who sows in May
Gets little that way. *January and May sowing.*

January commits the fault and May bears the blame.
[Applied in metaphor to human affairs also.] *January and May.*

A warm January, a cold May.

Morning red, foul weather and great need. *Jan. 1st.*

The first three days of January rule the coming three months. *1st, 2nd, 3rd*

Jan. 2nd. As the weather is this day, so will it be in September.

3rd. It will be the same weather for nine weeks as it is on the ninth day after Christmas.—SWEDEN.

6th. At twelfth day, the days are lengthened a cock's stride.
ITALIAN.

12th. If on January 12th the sun shine, it foreshows much wind.
SHEPHERD'S ALMANACK, 1676.

14th.
January 14th, St. Hilary,
The coldest day of the year.—YORKSHIRE.

22nd (St. Vincent's Day). If the sun shine brightly on Vincent's Day, we shall have more wine than water.—FRENCH.

Remember on St. Vincent's Day,
If that the sun his beams display,
Be sure to mark his transient beam,
Which through the casement sheds a gleam;
For 'tis a token bright and clear
Of prosperous weather all the year.

St. Vincent opens the seed.—SPANISH.

At St. Vincent all water is good as seed.—SPANISH.

If the sun shine on January 22nd, there shall be much wind.
HUSBANDMAN'S PRACTICE.

On St. Vincent's Day the vine sap rises to the branch, but retires frightened if it find frost.—FRENCH.

22nd and 25th.
If St. Vincent's has sunshine,
One hopes much rye and wine;
If St. Paul's is bright and clear,
One does hope a good year.

25th (St. Paul's Day).
St. Paul fair with sunshine
Brings fertility to rye and wine.

Fair on St. Paul's conversion day is favourable to all fruits.

If St. Paul's Day be faire and cleare,
It doth betide a happy yeare;
But if by chance it then should rain,
It will make deare all kinds of graine;
And if ye clouds make dark ye skie,
Then neate and fowles this yeare shall die;
If blustering winds do blow aloft,
Then wars shall trouble ye realm full oft.

If St. Paul's Day be fine, the year will be the same.—FRENCH.
This festival was called an Egyptian day; because (says Ducange) the Egyptians discovered that there were two unlucky days in every month, and prognostications of the good or bad course of the year were formed from the state of the weather on these days.

If St. Paul's Day be fair and clear, it indicates plenty; if cloudy or misty, much cattle will die; if rain and snow fall that day, it presages a dearth; if windy, it forebodes wars, as old wives do dream.—NATURE'S SECRETS (WILLSFORD). *Jan. 25th.*

If the sun shine on St. Paul's Day, it betokens a good year; if rain or snow, indifferent; if misty, it predicts great dearth; if thunder, great winds and death of people that year.

SHEPHERD'S ALMANACK, 1676.

The last twelve days of January rule the weather for the whole year. *19th to 31st.*

Hazel in first flower, January 31st; earliest in twenty years, January 15th.—MR. EDWARD MAWLEY. *31st.*

FEBRUARY.

Februeer
Both cut and shear. *Cold.*

Double-faced February. *Two-faced.*

Mad February takes his father into the sunshine and beats him.—SPANISH. *Mad.*

There is always one fine week in February. *Fine.*

All the months in the year
Curse a fair Februeer. *Fair.*

The Welshman had rather see his dam on the bier,
Than to see a fair Februeer.

When gnats dance in February, the husbandman becomes a beggar.

February, an ye be fair,
The hoggs 'll mend, and naething pair [lessen].
February, an ye be foul,
The hoggs 'll die in ilka pool.—TWEEDSIDE.

[Hoggs are sheep which have not been shorn.]

Isolated fine days in February are known in Surrey as "weather-breeders," and are considered as certain to be followed by a storm.

February singing,
Never stints stinging.

If bees get out in February, the next day will be windy and rainy.—SURREY.

A February spring is not worth a pin.—CORNWALL.

If in February there be no rain,
'Tis neither good for hay nor grain. *Rain.*

SPANISH AND PORTUGUESE.

February rain is only good to fill ditches.—FRENCH.

February fill the dyke,
Weather either black or white.

February fill dyke
With what thou dost like.—TUSSER.

[February.]

Wet.

February fill dyke, be it black or be it white;
But if it be white, it's better to like.

February fill ditch,
Black or white [*i.e.*, rain or snow], don't care which;
If it be white,
It's the better to like.

February fill dyke;
March lick it out.

When it rains in February, it will be temperate all the year.
SPANISH.

When it rains in February, all the year suffers.

Snow.

If February give much snow,
A fine summer it doth foreshow.—FRENCH.

Fogs.

Fogs in February mean frosts in May.

There will be as many frosts in June as there are fogs in February.

Thunder.

For every thunder with rain in February there will be a cold spell in May.

In February if thou hearest thunder,
Thou wilt see a summer's wonder.

Thunder in February or March, poor sugar [maple] year.

February, March, April, and May.

A dusty March, a snowy February, a moist April, and a dry May presage a good year.—FRENCH.

February and March.

When the cat in February lies in the sun, she will creep behind the stove in March. When the north wind does not blow in February, it will surely come in March.

February makes a bridge, and March breaks it.—T. FULLER.

February winds.

Violent north winds in February herald a fertile year.

Feb. 2nd.

Foul weather is no news;
Hail, rain, and snow
Are now expected, and
Esteemed no woe;
Nay, 'tis an omen bad,
The yeomen say,
If Phœbus shows his face
The second day.
COUNTRY ALMANACK FOR 1676.

On the eve of Candlemas Day
Winter gets stronger or passes away.—FRENCH.

Snow at Candlemas
Stops to handle us.—RUTLAND.

Feb. 2nd.

At Candlemas
Cold comes to us.

Candlemas Day! Candlemas Day!
Half our fire and half our hay!

[That is, we are midway through winter, and ought to have half our fuel and hay in stock.]

On Candlemas Day
You must have half your straw and half your hay.

Candlemas brings great pains.—FRENCH.

At Candlemas Day
Another winter is on his way.—FRENCH.

If Candlemas Day be fine and clear,
Corn and fruits will then be dear.

If Marie's purifying daie,
Be cleare and bright with sunnie raie,
Then frost and cold shall be much more
After the feast than was before.—A. FLEMING.

If Candlemas Day be fair and clear,
There'll be twa winters in the year.—SCOTCH.

You should on Candlemas Day
Throw candle and candlestick away.

As far as the sun shines in on Candlemas Day,
So far will the snow blow in afore old May.

The hind had as lief see his wife on the bier,
As that Candlemas Day should be pleasant and clear.

The shepherd would rather see the wolf enter his fold on Candlemas Day than the sun.

Should the sun shine out at the Purification (or churching of the Virgin Mary), there will be more ice after the festival than there was before it.—FROM THE LATIN PROVERB (SIR T. BROWNE'S "VULGAR ERRORS").

When on the Purification the sun hath shined,
The greater part of winter comes behind.

As far as the sun shines in at the window on Candlemas Day, so deep will the snow be ere winter is gone.

On Candlemas Day, just so far as the sun shines in, just so far will the snow blow in.

If Candlemas Day be fair and bright,
Winter will have another flight.
But if Candlemas Day bring clouds and rain,
Winter is gone and won't come again.

Feb. 2nd.

February 2nd, bright and clear,
Gives a good flax year.

If Candlemas Day be dry and fair,
The half of the winter's to come and mair.
If Candlemas Day be wet and foul,
The half of the winter is gone at Yule [Christmas].—SCOTCH.

After Candlemas Day the frost will be more keen,
If the sun then shines bright, than before it has been.

On Candlemas Day the bear, badger, or woodchuck comes out to see his shadow at noon: if he does not see it, he remains out; but if he does see it, he goes back to his hole for six weeks, and cold weather continues for six weeks longer.
UNITED STATES.

If the ground-hog is sunning himself on the 2nd, he will return for four weeks to his winter quarters again.

The badger peeps out of his hole on Candlemas Day, and when he finds snow walks abroad, but if he sees the sun shining he draws back into his hole.—GERMAN.

At the day of Candlemas,
Cold in air and snow on grass;
If the sun then entice the bear from his den,
He turns round thrice and gets back again.—FRENCH.

As long before Candlemas as the lark is heard to sing, so long will he be silent afterwards on account of the cold.
GERMAN.

Gif the lavrock sings afore Candelmas,
She'll mourn as lang after it.—SCOTCH.

As lang as the bird sings before Candlemas, it will greet after it.—SCOTCH.

On Candlemas Day, if the thorns hang a drop,
Then you are sure of a good pea crop.—SUSSEX.

[There is a similar proverb with respect to beans.]

If a storm on February 2nd, spring is near; but if that day be bright and clear, the spring will be late.

If it snows on February 2nd, only so much as may be seen on a black ox, then summer will come soon.

If on February 2nd the goose find it wet, then the sheep will have grass on March 25th.

When drops hang on the fence on February 2nd, icicles will hang there on March 25th.

When the wind's in the east on Candlemas Day,
There it will stick till the 2nd of May.

When it rains at Candlemas, the cold is over.—SPANISH.

When Candlemas Day is come and gone,
The snow lies on a hot stone.

Candlemas Day: Purification of the Virgin Mary.—The snowdrop, which was appropriately called "The fair maid of February," ought to blossom about this time. *Feb. 2nd.*

Sow or set beans in Candlemas waddle.*

St. Dorothea gives the most snow. *6th.*

If the eighteen last days of February be
Wet, and the first ten of March, you'll see
That the spring quarter, and the summer too
Will prove too wet, and danger to ensue. *10th to 28th.*

These three days, according to a Highland superstition, were said to be borrowed from January, and it is accounted a good omen if these days should be as stormy as possible. *12th to 14th.*

If the sun smile on St. Eulalie's Day,
It is good for apples and cider, they say.
FRENCH. *12th (St. Eulalie's Day).*

To St. Valentine the spring is a neighbour.—FRENCH. *14th (St. Valentine's Day).*

The crocus was dedicated to St. Valentine, and ought to blossom about this time.—CIRCLE OF THE SEASONS.

St. Valentine,
Set thy hopper † by mine.

Winter's back breaks about the middle of February.

The nights of this part of February are called in Sweden "steel nights," on account of their cutting severity. *20th to 28th.*

If cold at St. Peter's Day, it will last longer. *22nd (St. Peter's Day).*

The night of St. Peter shows what weather we shall have for the next forty days.

St. Matthias,
Sow both leaf and grass. *24th (St. Matthias' Day).*

If it freezes on St. Matthias' Day, it will freeze for a month together.

St. Matthias breaks the ice; if he finds none, he will make it.

St. Matthy
All the year goes by.

At St. Mattho
Take thy hopper † and sow.

St. Matthie
Sends sap into the tree.

The fair of Auld Deer [third Thursday in February]
Is the warst day in a' the year.—ABERDEEN.

* Wane of the moon. † Seed basket.

Feb. 28th.

Romanus bright and clear
Indicates a goodly year.

MARCH.

March, many weathers.

March many weathers rained and blowed,
But March grass never did good.—T. FULLER.

March yeans the lammie
And buds the thorn,
And blows through the flint
Of an ox's horn.—NORTHUMBERLAND.

In beginning or in end
March its gifts will send.

March was so angry with an old woman (according to a saying in the island of Kythnos) for thinking he was a summer month, that he borrowed a day from his brother February, and froze her and her flocks to death.—T. BENT (GREECE).

Dry.

Dust in March brings grass and foliage.

A dry and cold March never begs its bread.

A peck of March dust and a shower in May
Make the corn green and the fields gay.

March dust and March win'
Bleach as well as simmer's sin.—SCOTLAND.

A peck of March dust is worth a king's ransom.

A bushel of March dust on the leaves is worth a king's ransom.—T. FULLER.

A load of March dust is worth a ducat.—GERMAN.

A bushel of March dust is a thing
Worth the ransom of a king.

A March without water
Dowers the hind's daughter.—FRENCH.

Mild.

March flowers
Make no summer bowers.

Flies.

When flies swarm in March, sheep come to their death.
DUTCH.

Gnats.

When gnats dance in March, it brings death to sheep.—DUTCH.

Sun.

The March sun raises, but dissolves not.—G. HERBERT.

March sun
Lets snow stand on a stone.

The March sun wounds.—SPANISH.

March sun strikes like a hammer.—SPANISH.

Worse than the sun in March, *[March.] Sun.*
This praise doth nourish agues.
SHAKESPEARE'S "HENRY IV."

A March sun sticks like a lock of wool.

A wet March makes a sad harvest. *Rain*

March rain spoils more than clothes.

March wet and windy
Makes the barn full and finnie.—SCOTCH.

["Finnie" is used obliquely. The word means, in Scotland, the "feel" of the grain as indicating quality. This proverb is more generally applied to May: see p. 26.—P. DUDGEON.]

March damp and warm *Wet and warm.*
Will do farmer much harm.

March water is worse than a stain in cloth.

A March wisher [or whisher] *Fishing.*
Is never a good fisher.

March wind *Wind.*
Wakes the ether [adder] and blooms the whin.—SCOTLAND.

March mist, *Mist.*
Water in fist.—SPANISH.

So many mists in March you see,
So many frosts in May will be.

As many mistises in March,
So many frostises in May.—WILTSHIRE.

So many frosts in March, so many in May. *Frosts.*

A damp, rotten March gives pain to farmers. *Damp.*

As much dew in March, so much fog rises in August. *Dew.*

Snow in March is bad for fruit and grape vine. *Snow.*

In March much snow,
To plants and trees much woe.—GERMANY.

Fog in March, thunder in July. *Fog.*

As much fog in March, so much rain in summer.

Thunder in March betokens a fruitful year.—GERMAN. *Thunder.*

When it thunders in March, it brings sorrow.

When March thunders, tools and arms get rusty.
PORTUGUESE.

When it thunders in March, we may cry "Alas!"—FRENCH.

March, black ram,* *Stormy.*
Comes in like a lion and goes out like a lamb.

* An obscure expression [Aries ?], sometimes "balkham," "back ham," or "hack ham."

[March.] Stormy.

March comes in like a lamb and goes out like a lion.
[Reverse of the usual proverb.]

March comes in with adders' heads and goes out with peacocks' tails.—SCOTCH.

Cuckoo.

The cuckoo comes in mid March, and cucks in mid April;
And goes away at Lammas-tide, when the corn begins to fill.

Pruning.

He who freely lops in March will get his lap full of fruit.
PORTUGUESE.

Humours.

As Mars hasteneth all the humours feel it.

March and April.

When March has April weather, April will have March weather.—FRENCH.

March flings [kicks], April fleyes [warms].—SCOTCH.

March, April and May.

A windy March and a rainy April make a beautiful May.

March and May.

March wind and May sun
Make clothes white and maids dun.

Mists in March bring rain,
Or in May frosts again.

March, April and June.

March rainy, April windy, and then June will come beautiful with flowers.—SPANISH.

March, April and May.

March search, April try;
May will prove if you live or die.

March winds and April showers
Bring forth May flowers.

A dusty March, a snowy February, a moist April, and a dry May presage a good year.—FRENCH.

A dry March, wet April, and cool May
Fill barn, cellar, and bring much hay.

March and June.

As it rains in March, so it rains in June.

March and other months.

A frosty winter and a dusty March, and a rain about Averil,
Another about the Lammas time, when the corn begins to fill,
Is weel worth a pleuch [plough] o' gowd, and a' her pins theretill.
G. BUCHANAN.

March 1st (St. David's Day).

Upon St. David's Day
Put oats and barley in the clay.

1st and 2nd.

St. David and Chad,
Sow pease good or bad.

March 1st, 2nd, and 3rd.

First comes David, then comes Chad,
And then comes Winneral as though he was mad.
White or black,
Or old house thack.

[*Note.*—Meaning snow, rain, or wind—the latter endangering the thack or thatch.]

10th.

If it does not freeze on the 10th, a fertile year may be expected.

Mists or hoar frosts on this day betoken a plentiful year, but not without some diseases.

15th.

On March 15th come sun and swallow.—SPANISH.

17th (St. Patrick's Day).

St. Patrick's Day, the warm side of a stone turns up, and the broad-back goose begins to lay.

19th (St. Joseph's Day).

Is't on St. Joseph's Day clear,
So follows a fertile year.

21st (St. Benedict's Day).

St. Benedict,
Sow thy pease or keep them in thy rick.

When there has been no particular storm about the time of the spring equinox, if a storm arise from the east on or before that day, or if a storm from any point of the compass arise near a week after the equinox, then, in either of these cases, the succeeding summer is generally *dry*, four times in five; but if a storm arise from the S.W. or W.S.W. on or just before the spring equinox, then the summer following is generally *wet*, five times in six.—DR. KIRWAN.

25th (Lady Day).

Is't on St. Mary's bright and clear,
Fertile is said to be the year.

The flower cardamine, or lady's-smock, with its milk-white flowers, is dedicated to the Virgin Mary, and appears about Lady Day.

Borrowed days.

The three last days of March (old style) are called the borrowing days; for as they are remarked to be unusually stormy, it is feigned that March had borrowed them from April to extend the sphere of his rougher sway.—SIR W. SCOTT.

March borrowit from April
Three days, and they were ill:
The first was frost, the second was snaw,
The third was cauld as ever't could blaw.
SCOTCH.

March borrows of April
Three days, and they are ill;
April borrows of March again
Three days of wind and rain.

The warst blast comes in the borrowing days.

[March.] Borrowed days.

The Spanish story about the borrowing days is that a shepherd promised March a lamb if he would temper the winds to suit his flocks; but after gaining his point, the shepherd refused to pay over the lamb. In revenge March borrowed three days from April, in which fiercer winds than ever blew and punished the deceiver.

March borrowed of April, April borrowed of May,
Three days, they say:
One rained, and one snew,
And the other was the worst day that ever blew.
STAFFORDSHIRE.

The oldest North-Country version of the proverb about the borrowing days is the following:—

March said to Averil,
I see three hoggs [year-old sheep] on yonder hill;
An' if ye'll lend me dayis three,
I'll find a way to gar them dee.
The first o' them was wind an' weet;
The second o' them was snaw an' sleet;
The third o' them was sic' a freeze,
It froze the birds' nebs to the trees.
When the three days were past and gane,
The silly hoggs cam' hirplin hame.
SCOTLAND AND NORTH ENGLAND.

March borrowed from April
Three days, and they were ill:
The first of them is wan and weet,
The second it is snaw and sleet,
The third of them is a peel-a-bane,
And freezes the wee bird's neb to the stane.

Blackthorn winter.

There are generally some warm days at the end of March or beginning of April, which bring the blackthorn into bloom, and which are followed by a cold period called the "Blackthorn Winter."

APRIL.

A dry April
Not the farmer's will.
April wet
Is what he would get.

Rain.

In April each drop counts for a thousand.—SPANISH.

April rain is worth David's chariot.—FRENCH.

April showers bring summer flowers.

Flood.

An April flood carries away the frog and his brood.

In April Dove's * flood is worth a king's good.

* The river Dove in Derbyshire.

A cold April
The barn will fill.

[April.] Cold.

Cold April gives bread and wine.—FRENCH.

A cold April, much bread and little wine.—SPANISH.

April cold and wet fills barn and barrel.

Cold and wet.

A cold and moist April fills the cellar and fattens the cow.
PORTUGUESE.

A sharp April kills the pig.

April snow breeds grass.

Till April's dead
Change not a thread.

It is not April without a frosty crown.—FRENCH.

Frosty.

April wears a white hat.*

Changeable as an April day.

Change.

April weather
Rain and sunshine, both together.

Vine that buds in April
Will not the barrel fill.—FRENCH.

Buds.

Fogs in April foretell a failure of the wheat crop next year.
ALABAMA.

Fog.

You must look for grass in April on the top of an oak. Because the grass seldom springs well before the oak begins to put forth.—RAY.

Oak.

Plant your 'taturs when you will,
They won't come up before April.—WILTSHIRE.

Potatoes.

Whatever March does not want April brings along.

April and March.

Snow in April is manure; snow in March devours.

April and March snows.

A swarm of bees in April for me, and one in May for my brother.—SPAIN.

April and May.

In April much rain; in May a flood or two, and these not great.
SPAIN.

Betwixt April and May if there be rain,
'Tis worth more than oxen and wain.

Who ploughs in April ought not to have been born; who ploughs in May ought neither to have been born nor nursed.
SPANISH.

April and May are the keys of the year.

Milk of April and May.—SPANISH.

* Frost.

April and May. April and May between them make bread for all the year.
SPAIN.

April for me, May for my master.

Cloudy. Cloudy April, dewy May.—FRENCH.

Rain. April rains for men, May for beasts.
[*i.e.*, a rainy April is good for corn, and a wet May for grass crops.]

Let it rain in April and May for me,
And all the rest of the year for thee.—SPAIN.

April showers bring forth May flowers.

April and June. After a wet April a dry June.

Moist April, clear June.

April and autumn. The dews of April and May
Make August and September gay.—FRENCH.

After warm April and October, a warm year.

Thunder. Thunderstorm in April is the end of hoar-frost.

When April blows his horn,
It's good for hay and corn.

If it thunders on All Fools' Day
It brings good crops of corn and hay.

Early part of. The early part of April is called the blackthorn winter, because the thorn is then white with blossom and the weather generally cold.

First three days. If the first three days of April be foggy, there will be a flood in June.—HUNTINGDON.

April 3rd. The 3rd of April comes with the cuckoo and the nightingale.

6th (Latter Lady Day). On Lady Day the latter
The cold comes on the water.—T. FULLER.

14th. This day is called cuckoo day, and the cuckoo's song is generally first heard about this time.

Cuckoo. In Aprill, the koocoo can sing her song by rote;
In June, of tune she cannot sing a note:
At first, koo-coo, koo-coo, sing still can she do;
At last, kooke, kooke, kooke; six kookes to one koo.
HAYWOOD, 1587.

In April, come he will;
In May, he sings all day;
In June, he alters his tune;
In July, he prepares to fly;
In August, go he must.
If he stay till September,
'Tis as much as the oldest man can ever remember.

[April.]
Cuckoo.

The cuckoo in April,
He opens his bill;
The cuckoo in May,
He sings the whole day;
The cuckoo in June,
He changeth his tune;
The cuckoo in July,
Away he must fly.—NORTH YORKSHIRE.

In April, cuckoo sings her lay;
In May, she sings both night and day;
In June, she loses her sweet strain;
In July, she flies off again.—NORTH YORKSHIRE.

15th. This day is called Swallow Day, because swallows ought to appear at this date.

23rd (St. George's Day). If on St. George's Day the birch leaf is the size of a farthing, on the feast of our Lady of Kazan you will have corn in the barn.—RUSSIA.

When on St. George rye will hide a crow, a good harvest may be expected.

At St. George the meadow turns to hay.

23rd (St. George); 25th (St. Mark).

St. George cries "Goe!"
St. Mark cries "Hoe!"

As long before St. Mark's Day as the frogs are heard croaking, so long will they keep quiet afterwards.

MAY.

Merry.

The merry month of May.

Trust not a day
Ere birth of May.—LUTHER.

Hot.

A hot May makes a fat churchyard.

For a warm May
The parsons pray.
[Meaning more burial fees—a libellous proverb.]

Flowers.

Blossoms in May
Are not good, some say.

If May will be a gardener, he will not fill the granaries

Dry.

Dry May brings nothing gay.

Damp

May damp and cool fills the barns and wine vats.

Wet.

A May wet
Was never kind yet.

The haddocks are good
When dipped in May flood.

Rainy May marries peasants.—FRENCH.

Water in May is bread all the year.—SPAIN AND ITALY.

[May.] *Wet.*

A May flood
Never did good.

To be hoped for, like rain in May.—SPAIN.

Rain in the beginning of May is said to injure the wine.

A cold May is kindly,
And fills the barn finely.

A wet May
Makes a big load of hay.—WEST SHROPSHIRE.

A wet May
Will fill a byre full of hay.

May showers bring milk and meal.—SCOTCH.

A wet May and a winnie
Makes a fou stackyard and a finnie.—SCOTCH.

["Finnie"—the good quality, as judged by the *feel* of the corn.—P. DUDGEON.]

Cool and windy.

A cool May and a windy
Barn filleth up finely.—T. FULLER.

A cold May and a windy
Makes a barn full and a findy.

A cold May and a windy, a full barn will find ye.

[The three last are corrupt English versions of the Scotch proverb.]

A windy May makes a fair year.—PORTUGUESE.

A cold May is good for corn and hay.

Till May be out
Leave not off a clout.

OR—

Change not a clout
Till May be out.

May, come she early or come she late,
She'll make the cow to quake.—FRENCH.

Come it early or come it late,
In May comes the cow-quake [*i.e.*, tremulous grass].

Cold.

Cold May brings many things.

In the middle of May comes the tail of the winter.—FRENCH.

Cold May enriches no one.

Shear your sheep in May,
And shear them all away.

Dew. Cool and evening dew in May, brings wine and much hay.

Dry. For an east wind in May 'tis your duty to pray.

A snowstorm in May
Is worth a waggon-load of hay.

[*May.*]
Snowy.

Many thunderstorms in May,
And the farmer sings "Hey! hey!"

Thunder.

The more thunder in May, the less in August and September.

Be sure of hay till the end of May.—T. FULLER.

Hay.

In May much straw and little grain.—SPANISH.

To wed in May is to wed poverty.

Maids are May when they are maids; but the sky changes when they are wives.—SHAKESPEARE'S "AS YOU LIKE IT."

He who mows in May
Will have neither fruit nor hay.—PORTUGUESE.

Mowing.

He who sows oats in May
Gets little that way.

Sowing.

In May an east-lying field is worth wain and oxen; in June, the oxen and the yoke.

Be it weal or be it woe,
Beans blow before May doth go.

Beans.

Look at your corn in May,
And you will come weeping away;
Look at the same in June,
And you'll come home in another tune.

May and June.

[A proverb alluding to the magical way in which unpromising crops sometimes recover.]

The farmer went to his wheat in May,
And came sorrowing away;
The farmer went to his wheat in June,
And came away whistling a merry tune.—FRENCH.

A dry May is followed by a wet June

A dry May and a leaking June
Make the farmer whistle a merry tune.

They who bathe in May
Will soon be laid in clay;
They who bathe in June
Will sing a merry tune;
They who bathe in July
Will dance like a fly.

Mist in May, heat in June,
Make the harvest come right soon.

May and June.

A swarm of bees in May
Is worth a load of hay;
A swarm of bees in June
Is worth a silver spoon;
But a swarm in July
Is not worth a fly.

A misty May and a hot June
Bring cheap meal and harvest soon.

A leaking May and a warm June
Bring on the harvest very soon.—SCOTCH.

A leaky May and a dry June
Keep the poor man's head abune [above].
GREENOCK.

A dry May and a dripping June
Bring all things into tune.—BEDFORDSHIRE.

May and July.

Wet May, dry July.—GERMAN.

May and August.

Mud in May, grain in August.—SPANISH.

May and other months.

A red gay May, best in any year;
February full of snow is to the ground most dear;
A whistling March, that makes the ploughman blithe;
And moisty April, that fits him for the scythe.
WADRŒPHE, 1623.

May 1st.

Hoar-frost on May 1st indicates a good harvest.

The later the blackthorn in bloom after May 1st, the better the rye and harvest.

If it rains on Philip's and Jacob's Day, a fertile year may be expected.

8th.

If on the 8th of May it rain,
It foretells a wet harvest, men sain.—T. FULLER.

11th, 12th, and 13th.

St. Mamertius, St. Pancras, and St. Gervais do not pass without a frost.—FRANCE.

13th.

Who shears his sheep before St. Gervatius' day loves more his wool than his sheep.

25th.

At St. Urban gather your walnuts.—SPANISH.

JUNE.

Calm.

Calm weather in June
Sets corn in tune.

Fair.

It never clouds up in a June night for a rain.—UNITED STATES.

In the hay season, when there is no dew, it indicates rain. *[June.] Hay season.*

A cold and wet June spoils the rest of the year. *Wet.*

June damp and warm
Does the farmer no harm.

A good leak in June
Sets all in tune.

A dripping June
Brings all things in tune.

If north wind blows in June, good rye harvest. *North wind.*

In Scotland an early harvest is expected when the bramble blossoms early in June. *Harvest.*

When it is hottest in June, it will be coldest in the corresponding days of the next February. *June and February.*

A wet June makes a dry September.—CORNWALL. *June and September.*

If on the 8th of June it rain, *June 8th.*
It foretells a wet harvest, men sain.

If it rain on June 8th (St. Medard), it will rain forty days later; but if it rain on June 19th (St. Protais), it rains for forty days after.—FRENCH. *8th and 19th.*

On St. Barnabas *11th.*
Put a scythe to the grass.

Rain on St. Barnabas' Day good for grapes.

Barnaby bright,
The longest day and shortest night.

On St. Barnabas' Day
The sun is come to stay.—SPANISH.

If St. Vitus's Day be rainy weather, *15th.*
It will rain for thirty days together.

Oh! St. Vitus, do not rain, so that we may not want barley.

If it rains on midsummer eve, the filberts will be spoiled. *24th (St. John's Day).*

Before St. John's Day no early crops are worth praising.
GERMAN.

Before St. John's Day we pray for rain after that we get it anyhow.

Rain on St. John's Day, and we may expect a wet harvest.

Previous to St. John's Day we dare not praise barley.

If midsummer day be never so little rainy, the hazel and walnut will be scarce; corn smitten in many places; but apples, pears, and plums will not be hurt.—SHEPHERD'S KALENDAR.

June 24th (St. John's Day). Rain on St. John's Day, damage to nuts.

Cut your thistles before St. John,
You will have two instead of one.

Never rued the man
That laid in his fuel before St. John.—T. FULLER.

27th. If it rains on June 27th, it will rain seven weeks.

29th. If it rains on St. Peter's Day, the bakers will have to carry double flour and single water; if dry, they will carry single flour and double water.

Peter and Paul will rot the roots of the rye.

JULY.

July, God send thee calm and fayre,
That happy harvest we may see,
With quyet tyme and healthsome ayre,
And man to God may thankful bee.

Calm. No tempest, good July,
Lest corn come off blue by [mildew].

No tempest, good July,
Lest the corn look ruely.

Oysters. July, to whom, the dog-star in her train,
St. James gives oysters and St. Swithin rain.—CHURCHILL.

Sun in Leo. When the sun enters Leo, the greatest heat will then arise.

Sky. Ne'er trust a July sky.—SHETLAND.

Rye. In July
Shear your rye.

Rain. A shower of rain in July, when the corn begins to fill,
Is worth a plough of oxen, and all belongs theretill.

Thunder. Much thunder in July injures wheat and barley.

July and January. As July, so the next January.

July and August. In July
Some reap rye;
In August,
If one will not, the other must.

Whatever July and August do not boil, September cannot fry.

July, August and September. When the months of July, August, and September are unusually hot, January will be the coldest month.

July 1st. If the 1st of July it be rainy weather,
It will rain more or less for four weeks together.

2nd. If it rains on St. Mary's Day, it will rain for four weeks.

As the dog days commence, so they end. — *July 3rd to Aug.* 11*th (Dog days).*

If it rains on first dog day, it will rain for forty days after. — 3*rd.*

Dog days bright and clear
Indicate a happy year;
But when accompanied by rain,
For better times our hopes are vain.

If Bullion's Day be dry, there will be a good harvest.—SCOTCH. — 4*th.*
[St. Martin Bullion, to distinguish it from St. Martin's Day.—P. DUDGEON.]

Bullion's Day, gif ye be fair,
For forty days 'twill rain nae mair.—SCOTCH.

If the deer rise dry and lie down dry on Bullion's Day, there will be a good gose harvest.—SCOTCH.
["Gose," latter end of summer.]

If it rains on July 10th, it will rain for seven weeks. — 10*th.*

To the 12th of July from the 12th of May
All is day. — 12*th.*

If it rain on the feast of St. Processus and St. Martin, it suffocates the corn.—LATIN PROVERB, "NORWICH DOOMSDAY BOOK." — 14*th (O.S., July 2nd).*

Let not such vulgar tales debase thy mind,
Nor Paul nor Swithin rule the clouds and wind.—GAY. — 15*th (St. Swithin's Day)*

If about St. Swithin's Day a change of weather takes place, we are likely to have a spell of fine or wet weather.
C. W. EMPSON.

If St. Swithin weep, that year, the proverb says,
The weather will be foul for forty days.—T. FULLER.

If St. Swithin greets, the proverb says,
The weather will be foul for forty days.—SCOTCH.

In this month is St. Swithin's Day,
On which if, that it rain they say,
Full forty days after it will
Or more or less some rain distil.
POOR ROBIN'S ALMANACK, 1697.

St. Swithin is christening the apples.
[This saying is applied to rain on St. Swithin's Day.]

St. Swithin's Day, if ye do rain,
For forty days it will remain;
St. Swithin's Day, an' ye be fair,
For forty days 'twill rain nae mair.—SCOTCH.

July 15th. It is said in Tuscany that the weather on St. Gallo's Day (July 15th) will prevail for forty days; and at Rome the period is extended to any day within the octave of St. Bartholomew.

July 15th and August 24th.

All the tears that St. Swithin can cry
St. Bartlemy's dusty mantle wipes dry.

19th. At St. Vincent the rain ceases and the wind comes.—FRENCH.

20th.

Clear on St. Jacob's Day, plenty of fruit.

So much rain often falls about this day that people often speak of "Margaret's flood."

Rain on St. Margaret's Day will destroy all kinds of nuts.

GERMAN.

22nd Mary Magdalene's Day. The roses are said to begin to fade on this day.

Alluding to the wet usually prevalent about the middle of July, the saying is: "St. Mary Magdalene is washing her handkerchief to go to her cousin St. James's fair.

FOLK-LORE JOURNAL.

25th.

Till St. James's Day be come and gone,
You may have hops and you may have none.

AUGUST.

Dry.

Dry August and warm
Doth harvest no harm.

Sunshine. August sunshine and bright nights ripen the grapes.

Wet. August rain gives honey, wine, and saffron.—PORTUGUESE.

When it rains in August, it rains honey and wine.

FRENCH AND SPANISH.

A wet August never brings dearth.—ITALIAN.

Fogs.

So many August fogs, so many winter mists.

Observe on what day in August the first heavy fog occurs, and expect a hard frost on the same day in October.

UNITED STATES.

A fog in August indicates a severe winter and plenty of snow.

Dew. When the dew is heavy in August, the weather generally remains fair. Thunderstorms in the beginning of August will generally be followed by others all the month.

August and February.

As August, so the next February.

August and September.

August ripens, September gathers in;
August bears the burden, September the fruit.

PORTUGUESE.

August and December.

None in August should over the land,
In December none over the sea.

After Lammas corn ripens as much by night as by day. [*August.*] *Lammas Day.*

[*Note.*—Alluding to the heavy night dews.]

If the first week in August is unusually warm, the winter will be white and long. *First week.*

St. Margaret's flood is proverbial, and is considered to be well for the harvest in England. *Old Style, August 1st; New Style, August 13th.*

If on St. Lawrence's Day the weather be fine, fair autumn and good wine may be hoped for.—GERMAN. *10th.*

On St. Mary's Day sunshine
Brings much and good wine. *15th.*

If this day be misty, the morning beginning with a hoar-frost, the cold weather will soon come, and a hard winter. *24th (St. Bartholomew's Day).*

SHEPHERD'S KALENDAR.

If it rains this day, it will rain the forty days after.—ROMAN.

At St. Bartholomew
There comes cold dew.

St. Bartlemy's mantle wipes dry
All the tears that St. Swithin can cry.

If the 24th of August be fair and clear,
Then hope for a prosperous autumn that year.

As Bartholomew's Day, so the whole autumn.

Thunderstorms after Bartholomew's Day are generally violent.

SEPTEMBER

September dries up wells or breaks down bridges. *Dry or wet.*

PORTUGUESE.

'Tis September's sun which causes the black list upon the antelope's back.—BOMBAY. *Sun.*

As September, so the coming March. *September and March.*

A wet September, drought for next summer, famine, and no crops.—CALIFORNIA. *Wet.*

Heavy September rains bring drought.—UNITED STATES. *Rain.*

Rain in September is good for the farmer, but poison to the vine-growers.—GERMAN.

September rain is much liked by the farmer.

September rain good for crops and vines.

If the storms in September clear off warm, all the storms of the following winter will be warm. *Storms.*

[September.] Cold. When a cold spell occurs in September and passes without a frost, a frost will not occur until the same time in October.

Thunder. Thunder in September indicates a good crop of grain and fruit for next year.

Fodder. Preserve your fodder in September and your cow will fatten.
PORTUGUESE.

September and November.
September blow soft till the fruit's in the loft.
November take flail, let ships no more sail.

Sept. 1st. Fair on September 1st, fair for the month.

St. Giles finishes the walnuts.—SPANISH.

8th. As on the 8th, so for the next four weeks.

14th (Holyrood). The passion flower blossomed about this time. The flower is said to present a resemblance to the cross or rood, the nails, and the crown of thorns, used at the Crucifixion.
CIRCLE OF THE SEASONS.

If dry be the buck's horn
On Holyrood morn,
'Tis worth a kist of gold;
But if wet it be seen
Ere Holyrood e'en,
Bad harvest is foretold.—YORKSHIRE.

If the hart and the hind meet dry and part dry on Rood Day fair,
For sax weeks, of rain there'll be nae mair.—SCOTCH.

On Holy-Cross Day
Vineyards are gay.—SPANISH.

Three windy days. There are generally three consecutive windy days about the middle of September, which have been called by the Midland millers the windy days of barley harvest.

15th. This day is said to be fine in six years out of seven.
T. FORSTER'S "PERENNIAL CALENDAR."

19th. If on September 19th there is a storm from the south, a mild winter may be expected.—DERBY.

20th, 21st, and 22nd. These three days of September rule the weather for October, November, and December.

21st (St. Matthew's Day).
St. Matthee,
Shut up the bee.

St. Matthew's rain fattens pigs and goats.—SPAIN.

St. Matthew
Brings on the cold dew.

St. Matthew makes the days and nights equal.—SPANISH.

Sept. 21st (St. Matthew's Day).

Matthew's Day, bright and clear,
Brings good wine in next year.

South wind on September 21st indicates that the rest of the autumn will be warm.

St. Matthew and St. Matthias.

St. Matthew,
Get candlesticks new;
St. Mathi,
Lay candlesticks by.

29th (Michaelmas Day).

So many days old the moon is on Michaelmas Day, so many floods after.—HOWELL.

On Michaelmas Day the devil puts his foot on the blackberries.
NORTH OF IRELAND.

If St. Michael brings many acorns, Christmas will cover the fields with snow.

Michaelmas rot
Comes ne'er in the pot.

St. Michael's rain does not stay long in the sky.—FRENCH.

September 29th and October 16th.

If it does not rain on St. Michael's and Gallus, a dry spring is indicated for the next year.

OCTOBER.

Dry your barley in October,
Or you'll always be sober.
[Because if this is not done there will be no malt.—SWAINSON.]

Wind.

A good October and a good blast,
To blow the hog acorn and mast.

Fine

There are always nineteen fine days in October.—KENT.

Rain.

Much rain in October, much wind in December.

Cold.

When it freezes and snows in October, January will bring mild weather; but if it is thundering and heat-lightning, the weather will resemble April in temper.

Frosts, etc.

If October bring heavy frosts and winds, then will January and February be mild.

Snow.

If the first snow falls on moist, soft earth, it indicates a small harvest; but if upon hard, frozen soil, a good harvest.

Fogs.

For every fog in October a snow in the winter, heavy or light according as the fog is heavy or light.

Leaves.

If in the fall of the leaves in October many of them wither on the boughs and hang there, it betokens a frosty winter and much snow.

October and February.

Warm October, cold February.

October, January, and February.

If October bring much frost and wind, then are January and February mild.

October and March. As the weather in October, so will it be in the next March.

October and winter. When birds and badgers are fat in October, expect a cold winter. UNITED STATES.

Moon. Full moon in October without frost, no frost till full moon in November.

October and November. Plenty of rain in October and November on the North Pacific coast indicates a mild winter; little rain in these months will be followed by a severe winter.

Manure.

In October dung your field,
And your land its wealth shall yield.

October 18th.

St. Luke's little summer.

There is often about this time a spell of fine, dry weather, and this has received the name of St. Luke's little summer.

28th. (SS. Simon and Jude). This day was anciently accounted as certain to be rainy.

On St. Jude's Day
The oxen may play.

NOVEMBER.

Windy.

November take flail,
Let ships no more sail.—TUSSER.

Cheerless.

No warmth, no cheerfulness, no healthful ease,
No comfortable feel in any member,
No shade, no shine, no butterflies, no bees,
No fruits, no flowers, no leaves, no birds—
No-vember.—T. HOOD.

Flowers. Flowers in bloom late in autumn indicate a bad winter.

Water. When in November the water rises, it will show itself the whole winter.

Cold.

If there's ice in November that will bear a duck,
There'll be nothing after but sludge and muck.

A heavy November snow will last till April.—NEW ENGLAND.

Thunder.

Thunder in November, a fertile year to come.

Thunder in November on the Northern lakes is taken as an indication that the lakes will remain open till at least the middle of December.—UNITED STATES.

November and March.

As November, so the following March.

Nov. 1st (All Saints' Day).

On the 1st of November, if the weather hold clear,
An end of wheat sowing do make for the year.

In Sweden there is often about this time some warm weather, called "The All Saints' rest."

In Shakespeare's *Henry IV.*, Act I., Scene 2, mention is also made of the All Hallow'n summer.—SWAINSON.

Nov. 1st (All Saints' Day).

If All Saints' Day will bring out the winter, St. Martin's Day will bring out Indian summer.—UNITED STATES.

If on All Saints' Day the beech nut is dry, we shall have a hard winter; but if the nut be wet and not light, we may expect a wet winter.

11th (St. Martin's Day).

If ducks do slide at Hollantide,
At Christmas they will swim;
If ducks do swim at Hollantide,
At Christmas they will slide.

If it is at Martinmas fair, dry, and cold, the cold in winter will not last long.

If the geese at Martin's Day stand on ice, they will walk in mud at Christmas.

If the leaves of the trees and grape vines do not fall before Martin's Day, a cold winter may be expected.

When the wind is in this quarter (S.S.W.) at Martinmas, it keeps mainly to the same point right on to Old Candlemas Day (February 14th), and we shall have a mild winter up to then and no snow to speak of.—VERIFIED IN 1869 (SEE "NOTES AND QUERIES," MAY 8TH, 1869).

Wind north-west at Martinmas, severe winter to come.
HUNTINGDONSHIRE.

If the wind is in the south-west at Martinmas, it keeps there till after Candlemas, with a mild winter up to then and no snow to speak of.—MIDLAND COUNTIES.

At St. Martin's Day
Winter is on his way.—FRENCH.

Expect St. Martin's summer, halcyon days [*i.e.*, fine weather at Martinmas].—SHAKESPEARE'S "HENRY VI.," PART I., ACT. I., SCENE 2.

Weather folk-lore.

It is an old saying with the people round here (Atherstone), "Where the wind is on Martinmas Eve, there it will be the rest of the winter." The following, from Brand's *Popular Antiquities*, has reference to the first part of the foregoing: "The weather on Martinmas Eve is anxiously watched by the farmers in the Midland Counties, as it is supposed to be an index to the barometer for some two or three months forward."

November 11th and December 25th.

'Tween Martinmas and Yule
Water's wine in every pool.—SCOTLAND.

Nov. 21st.

As November 21st, so is the winter.

25th.

As at Catherine foul or fair, so will be the next February.

DECEMBER.

Cold. December cold with snow, good for rye.

Thunder. Thunder in December presages fine weather.

December and January.

December's frost and January's flood
Never boded the husbandman's good.

1st Sunday. If it rains on this Sunday before Mass, it will rain for a week.

Dec. 11th, Halcyon days. The fourteen halcyon days then began—days in which in the Mediterranean a calm weather was expected, so that the halcyon or hawk could (it was supposed) make its nest on the surface of the sea.—SEE VIRGIL'S "GEORGICS," BOOK I., LINE 393.

21st (St. Thomas's Day). Look at the weathercock on St. Thomas's Day at twelve o'clock, and see which way the wind is, for there it will stick for the next (lunar) quarter.

25th. A green Christmas makes a fat churchyard.

A green Christmas brings a heavy harvest.—RUTLAND.

At Christmas meadows green, at Easter covered with frost.

A clear and bright sun on Christmas Day fortelleth a peaceable year and plenty; but if the wind grow stormy before sunset, it betokeneth sickness in the spring and autumn quarters.

Christmas sunshine. The shepherd would rather see his wife enter the stable on Christmas Day than the sun.—GERMAN.

If the sun shine through the apple tree on Christmas Day, there will be an abundant crop in the following year.

Light Christmas,* light wheatsheaf;
Dark Christmas, heavy wheatsheaf.

Windy. If windy on Christmas Day, trees will bring much fruit.

Christmas and Easter.

A warm Christmas, a cold Easter;
A green Christmas, a white Easter.—GERMAN.

Easter in snow, Christmas in mud;
Christmas in snow, Easter in mud.

Wet. Christmas wet, empty granary and barrel.

Ice and snow. If Christmas finds a bridge, he'll break it; if he finds none he'll make one.

Snow. If it snows during Christmas night, the crops will do well.

So far as the sun shines on Christmas Day,
So far will the snow blow in May.—GERMANY.

Snow on Christmas night, good hop crop next year.

* If full moon about Christmas Day.

If at Christmas ice hangs on the willow, clover may be cut at Easter. *Dec. 25th. Ice.*

If ice will bear a man before Christmas, it will not bear a mouse afterwards. [Also said of a goose and duck.]

When the blackbird sings before Christmas, she will cry before Candlemas.—MEATH. *Blackbird.*

Superstitious rhyme.

> If Christmas Day on Thursday be,
> A windy winter ye shall see;
> Windy weather in each week,
> And hard tempest strong and thick,
> The summer shall be good and dry,
> Corn and beasts shall multiply;
> The year is good for lands to till,
> Kings and princes shall die by skill, etc., etc.

[There are eight more lines of the same superstitious character, but not relating to the weather.]

A windy Christmas and a calm Candlemas are signs of a good year. *Christmas and Candlemas.*

If on Christmas night the wine ferments heavily in the barrels, a good wine year is to follow. *Wine.*

Thunder during Christmas week indicates that there will be much snow during the winter. *Thunder.*

These twelve days are said to be the keys of the weather for the whole year. *Dec. 25th to Jan. 5th.*

There was a superstitious practice in France on Christmas Day of placing twelve onions, representing the twelve months. Each onion had a pinch of salt on the top; and if the salt had melted by Epiphany, the month corresponding was put down as sure to be wet; while if the salt remained, the month was to be dry. *Christmas to Epiphany.*

If it rain much during the twelve days after Christmas, it will be a wet year.

St. Stephen's Day windy, bad for next year's grapes. *Dec. 26th.*

The *Shepherd's Kalendar* mentions that if it be lowering and wet on Childermas Day, there will be scarcity; while if the day be fair, it promises plenty. *28th (Innocents' Day).*

31st.

> If New Year's Eve night wind blow south,
> It betokeneth warmth and growth;
> If west, much milk and fish in the sea;
> If north, much cold and storms there will be;
> If east, the trees will bear much fruit;
> If north-east, flee it man and brute.

EQUINOX. As the wind and weather at the equinoxes, so will they be for the next three months.

As the equinoctial storms clear, so will all storms clear for the six months.

Vernal equinox, wind N.E. and S.W. Wind north-east or north at noon of the vernal equinox, no fineweather before midsummer. If westerly or south-westerly, fine weather till midsummer.

If the wind is north-east at vernal equinox, it will be a good season for wheat and a poor one for other kinds of corn; but if south or south-west, it will be good for other corn, but bad for wheat.

Equinoctial gales. The vernal equinoctial gales are stronger than the autumnal.

If near the time of the equinox it blows in the day, it generally hushes towards evening.

PROVERBS RELATING TO VARIOUS MOVABLE FEASTS, ETC.

Shrove Tuesday. So much as the sun shineth on Pancake Tuesday, the like will shine every day in Lent.

Thunder on Shrove Tuesday foretelleth wind, store of fruit, and plenty.

When the sun is shining on Shrovetide Day, it is meant well for rye and peas.

Ash Wednesday. Wherever the wind lies on Ash Wednesday, it continues during all Lent.

As Ash Wednesday, so the fasting time.

Lent. Dry Lent, fertile year.

Palm Sunday. If the weather is not clear on Palm Sunday, it means a bad year.

Good Friday. Rain on Good Friday foreshows a fruitful year.

Good Friday and Easter Day.

A wet Good Friday and a wet Easter Day
Make plenty of grass, but very little hay.

Easter. Late Easter, long, cold spring.—SUSSEX.

Rain at Easter gives slim fodder.—UNITED STATES.

A rainy Easter betokens a good harvest.—FRENCH.

If the sun shines on Easter Day, it shines on Whitsunday likewise. *[Easter.]*

Frost.

Past Easter frost,
Fruit not lost.

Rain.

A good deal of rain upon Easter Day
Gives a good crop of grass, but little good hay.
HERTFORDSHIRE.

Weather.

Such weather as there is on Easter Day there will be at harvest.

[As a correspondent in *Notes and Queries* (July 10th, 1875) points out, this superstition may have arisen from the pagan sacrifice to the goddess Eostre (from which name the Venerable Bede says "Easter" is derived), a sacrifice made about the vernal equinox, with a view to a good harvest.]

First Sunday after.

The first Sunday after Easter settles the weather for the whole summer.—SWEDEN.

Pastor Sunday (second after Easter).

If it rains on Pastor Sunday, it will rain every Sunday until Pentecost (Whitsunday).

Holy Thursday.

Fine on Holy Thursday, wet on Whit-Monday; fine on Whit-Monday, wet on Holy Thursday.—HUNTINGDONSHIRE.

Ascension Day.

As the weather on Ascension Day, so may be the entire autumn.

Easter to Whitsuntide.

If fair weather from Easter to Whitsuntide, the butter will be cheap.

Corpus Christi (Thursday after Trinity Sunday).

Corpus Christi Day clear
Gives a good year.

If it rain on Corpus Christi Day, the rye granary will be light.

Whitsuntide.

Whitsuntide rain, blessing for wine.

Rain at Whitsuntide is said to make the wheat mildewed.

Strawberries at Whitsuntide indicate good wine.

Whitsunday (called also Pentecost—the fiftieth day after Easter).

Whitsunday bright and clear
Will bring a fertile year.

If Whitsunday bring rain, we expect many a plague.

Rain at Pentecost forebodes evil.

Whitsunday and Christmas.

Whitsunday wet, Christmas fat.

PROVERBS RELATING TO THE MONTHS GENERALLY.

Month. The month that comes in good will go out bad.

Satire.
Dirty days hath September,
April, June, and November;
From January up to May,
The rain it raineth every day.
All the rest have thirty-one,
Without a blessed gleam of sun;
And if any of them had two-and-thirty,
They'd be just as wet and twice as dirty.—MAINE, U.S.

Character of.
January fierce, cold, and frosty,
February moist and aguish,
March dusty,
April rainy,
May pretty, gay, and windy,
Bring an abundant harvest.—FRENCH.

A frosty winter and a dusty March,
And a rain about Aperill,
And another about the Lammas * time,
When the corn begins to fill,
Is worth a plough of gold
And all her pins theretill.

[A Scotch version of this, attributed to G. Buchanan, will be found among the March proverbs, p. 20.]

DAYS OF THE WEEK.

These sayings, though, for the most part, purely superstitious, I have inserted in order to complete the collection.

Wednesday clear. When the sun sets clear on Wednesday, expect clear weather the rest of the week.

Wednesday clearing, clear till Sunday.

Thursday.
On Thursday at three
Look out, and you'll see
What Friday will be.—SOUTH DEVON.

Friday.
Friday's a day as'll have his trick,
The fairest or foulest day o' the wik [week].
SHROPSHIRE.

Friday and Sunday.
Fine on Friday,
Fine on Sunday;
Wet on Friday,
Wet on Sunday.—FRANCE.

As the Friday, so the Sunday.

* August 1st.

If on Friday it rain,
'Twill on Sunday again;
If Friday be clear,
Have for Sunday no fear.

Friday and Sunday.

If the sun sets clear on Friday, it will blow before Sunday night.

Right as the Friday sothly for to tell
Now shineth it and now it raineth fast;
Right so can gery Venus overcast
The hertes of hire folk, right as her day
Is gerfull, right so changeth she aray;
Selde is the Friday all the weke ylike.

CHAUCER'S "KNIGHT'S TALE."

Friday's weather.

Friday is the best or worst day of the week.

If the sun sets clear on Friday, generally expect rain before Monday.

There is never a Saturday without some sunshine. *Saturday.*

If it rains on Sunday before Mass, it will rain all the week. *Sunday.*

Sunday clearing, clear till Wednesday. *Clearing.*

If sunset on Sunday is cloudy, it will rain before Wednesday. *Sunset.*

When it storms on the first Sunday in the month, it will storm every Sunday. *First in month.*

The last Sunday in the month indicates the weather of the next month. *Last in month.*

A misty morning may have a fine day.—T. FULLER. *Day misty.*

Too bright a morning breeds a lowering day.

PLAY OF "EDWARD III."

When there are three days cold, expect three days colder. *Cold.*

A warm and serene day, which we say is too fine for the season, betokens a speedy reverse.—"WHITBY GLOSSARY," F. K. ROBINSON. *Fine.*

Frosty nights and hot sunny days
Set the corn fields all in a blaze.

Days and nights.

A bad day has a good night. *Day and night.*

A day should be praised at night.—NORWAY.

If a change of weather occur when the sun or moon is crossing the meridian, it is for twelve hours at least.

NAUTICAL.

Noon change.

Twilight looming indicates rain. *Twilight.*

In the evening one may praise the day.—GERMAN. *Evening.*

If the weather change at night, it will not last when the day breaks.—FRANCE. *Night.*

Hours.
10 and 2.

Between the hours of ten and two
Will show you what the day will do.

12 and 2.

Between twelve and two
You'll see what the day will do.—CORNWALL.

7 and 11.

Rain at seven, fine at eleven;
Rain at eight, not fine till eight.

Cycle of change.

Lord Bacon states that it is an old opinion that the weather changes after forty years repeat themselves.

[*Note.*—The closest observation in modern times has failed to fix any period after which the weather may be said to repeat its changes.]

LIST OF COMMON PLANTS,

And the dates at which they ought to be in full flower. The forwardness of the seasons may be judged by the punctuality of the appearance of the blossoms.

Jan.	2	Groundsel
,,	4	Hazel
,,	5	Bearsfoot
,,	6	Common Dead Nettle
,,	9	Laurel
,,	10	Gorse
,,	11	Early Moss
,,	14	Barren Strawberry
,,	15	Ivy
,,	17	Anemone (Garden)
,,	19	White Dead Nettle
,,	27	Earth Moss
,,	28	Double Daisy
,,	30	Maidenhair
Feb.	1	Bay
,,	2	Snowdrop
,,	5	Primrose
,,	6	Blue Hyacinth
,,	9	Narcissus (Roman)
,,	13	Polyanthus
,,	14	Yellow Crocus
,,	17	Scotch Crocus
,,	19	Speedwell
,,	21	White Crocus
,,	22	Common Daisy
,,	23	Apricot
,,	25	Peach
,,	26	Periwinkle (Lesser)
,,	28	Purple Crocus
March	1	Leek
,,	4	Chickweed
,,	5	Hellebore
,,	6	Lent Lily
,,	7	Early Daffodil
,,	8	Great Jonquil
,,	13	Heartsease
,,	15	Coltsfoot
,,	17	Shamrock
,,	17	Violet
,,	24	Saxifrage
,,	25	Marigold
,,	29	Oxlip
,,	30	Cardamine
,,	30	Lesser Daffodil
April	2	White Violet
,,	4	Crown Imperial
,,	7	Anemone (Wood)
,,	8	Ground Ivy
,,	9	Polyanthus (Red)
,,	11	Dandelion
,,	12	Saxifrage (Great)
,,	13	Narcissus (Green)
,,	16	Yellow Tulip
,,	19	Garlic
,,	23	Harebell
,,	24	Blackthorn
,,	27	Great Daffodil
,,	30	Cowslip

May 2 Charlock
„ 2 Rhododendron
„ 3 Narcissus (Poetic)
„ 5 Apple Tree
„ 8 Lily of the Valley
„ 9 Solomon's Seal
„ 11 Asphodel (Yellow)
„ 14 Common Peony
„ 16 Star of Bethlehem
„ 17 Poppy (Early Red)
„ 18 Mouse Ear
„ 19 Monkshood
„ 20 Horse Chestnut
„ 23 Lilac
„ 26 Azalea (Yellow)
„ 27 Buttercup

June 1 Yellow Rose
„ 2 Pimpernel
„ 4 Pink (Indian)
„ 6 Pink (Common)
„ 8 Moneywort
„ 9 Barberry
„ 10 Fleur de Lis (Yellow)
„ 12 White Dog Rose
„ 13 Ranunculus (Garden)
„ 15 Sensitive Plant
„ 16 Moss Rose
„ 18 Poppy (Horned)
„ 22 Canterbury Bell
„ 23 Lady's Slipper
„ 24 St. John's Wort
„ 25 Sweet William
„ 26 Sow Thistle (Blue)
„ 28 Cornflower

July 1 Agrimony
„ 2 White Lily
„ 7 Nasturtium
„ 11 Yellow Lupin
„ 12 Snap Dragon
„ 13 Blue Lupin
„ 14 Red Lupin
„ 16 Convolvolus Major
July 17 Sweet Pea
„ 23 Musk Flower
„ 25 Herb Christopher
„ 26 Camomile (Field)

Aug. 2 Tiger Lily
„ 3 Hollyhock
„ 4 Bluebell
„ 6 Meadow Saffron
„ 7 Amaranth (Common)
„ 8 Love lies Bleeding
„ 10 Balsam (Common)
„ 11 China Aster
„ 12 Sow Thistle (Great)
„ 18 African Marigold
„ 21 Sunflower
„ 28 Golden Rod
„ 29 Yellow Hollyhock
„ 31 Pheasant's Eye

Sept. 5 Mushroom
„ 10 Autumnal Crocus
„ 14 Passion Flower

Oct. 2 Common Soapwort
„ 4 Southernwood
„ 5 Camomile (Starlike)
„ 6 Fever Few (Late Flowering)
„ 7 Crysanthemum (Indian)
„ 11 Holly
„ 16 Yarrow
„ 17 Sunflower (Ten Leaved)

Nov. 1 Laurestine
„ 6 Yew
„ 25 Butterbur (Sweet)

Dec. 4 Gooseberry (Barbadoes)
„ 7 Achania (Hairy)
„ 8 Arbor Vitæ
„ 23 Cedar of Lebanon
„ 26 Purple Heath

HONE'S "EVERY-DAY BOOK."

FLOWERS

Which should open on certain saints' days.

Feb.	2	Candlemas, Snowdrop
,,	14	St. Valentine, Crocus
March	25	Lady Day, Daffodil
April	23	St. George, Harebell
May	3	Holy Cross, Crowfoot
June	11	St. Barnabas, Ragged Robin
,,	24	St. John the Baptist, Scarlet Lychnis
July	15	St. Swithin, Lily
,,	20	St. Margaret, Poppy
July	22	St. Magdalene, Rose
Aug.	1	Lammas, Camomile
,,	15	Assumption, Virgin's Bower
,,	24	St. Bartholomew, Sun flower
Sept.	14	Holyrood, Passion Flower
,,	29	Michaelmas, Michaelmas Daisy
Nov.	25	St. Catherine, Laurel
Dec.	25	Christmas, Ivy and Holly

LIST OF COMMON FLOWERS,

And the times at which, in ordinary fine weather, they open and close their petals. Their opening later or closing earlier than the usual time is a sign of rain, and vice versâ.

	OPENS. A.M.	CLOSES. P.M.
Goatsbeard	3 to 5	9 to 10
Succory	4 ,, 5	8 ,, 9
Ox Tongue	4 ,, 5	12
Naked Poppy	5	7
Day Lily	5	7 ,, 8
Sow Thistle	5	11 ,, 12
Blue Thistle	5	12
Dandelion	5 ,, 6	8 ,, 9
Convolvulus	5 ,, 6	4 ,, 5
Spotted Hawkweed	6 ,, 7	4 ,, 5
Lettuce	7	10
White Water Lily	7	5
African Marigold	7	3 ,, 4
Pimpernel	7 ,, 8	2 ,, 3
Proliferous Pink	8	6
Mouse Ear	8	2
Field Marigold	9	3
Chickweed	9 ,, 10	9 ,, 10
Caroline Mallow	9 ,, 10	12 ,, 1

BIRDS,

And the times at which they usually appear in the South of England.

Wryneck	Middle of March.
Smallest Willow Wren	Latter end of March.
House Swallow	Middle of April.
Martin	,,
Sand Martin	,,
Blackcap	,,

Nightingale	Beginning of April.
Cuckoo	Middle of April.
Middle Willow Wren	"
Whitethroat	"
Redstart	"
Great Plover or Stone Curlew	End of March.
Grasshopper Lark	Middle of April.
Swift	Latter end of April.
Largest Willow Wren	End of April.
Fern Owl	Latter end of May.
Flycatcher	Middle of May.

T. FORSTER'S "PERENNIAL CALENDAR."

WINTER BIRDS.

Times of their arrival.

Ring Ouzel	Soon after Michaelmas.
Redwing	Middle of October.
Fieldfare	October and November.
Royston Crow	October.
Woodcock	Keeps arriving all October and November.
Snipe	The same (some of them breed here).
Jack Snipe	" " "
Pigeon or Stock Dove	End of November (some abide here all the year).
Wood Pigeon or Ring Dove	Some abide all the year; some arrive in spring; others perform partial migrations.

T. FORSTER'S "PERENNIAL CALENDAR."

Sun, Moon, and Stars.

The indications of coming weather presented by the sun, moon, etc., come next in order, and they refer for the most part to the weather of the day, or very soon after. The sun has ever been the first authority, and has his various aspects, colours, and moods, each fitted with a real or imaginary sequence of weather. His redness on rising or setting has furnished the material for a dozen proverbs of various times and nations. The moon, too, has always had her votaries as a weather witch, and even now is not without a numerous staff of prophets ready to assert her influence over the rain and clouds. One frequently hears of the weather altering at the "change of the moon," but careful observers have been unable to detect any real differences in the state of the air at such times. A more extended observation, however, will do the subject no harm, and may lead to the discovery of a law or the establishment of some rule on which reliance can be placed. The appearance of a halo round the moon is regarded as an indication of wet weather, and from its relative position gives some warning as to the time when the coming change may be expected.

SUN.

Red. A red sun has water in his eye.

Beams. When solar rays are visible in the air, they indicate vapour and rain to follow, and the sun is said to be "drawing water."

Rays. The pillars of light which are seen upright, and do commonly shoot and vary, are signs of cold; but both these are signs of drought.—BACON.

When the sun's rays are visible, the seamen say, "The sun is getting up his back stays, and it is time to look out for bad weather."

The sun breaking out suddenly into bright sunshine through an otherwise stormy sky is said to be making holes for the wind to blow through.—ROPER'S "WEATHER SAYINGS."

Clouds. The sun is noted to be hotter when it shineth forth between clouds, than when the sky is open and serene.—BACON.

Heat. The heat or beams of the sun doth take away the smell of flowers, specially such as are of milder odour.—BACON.

[SUNRISE.] If rays precede the sunrise, it is a sign both of wind and rain.
BACON.

Morning. The morning sun never lasts the day.

Halo. If the rising sun be encompassed with an iris or circle of white clouds, and they equally fly away, this is a sign of fair weather.
PLINY.

Concave. If the sun appear concave at its rising, the day will be windy or showery,—windy if the sun be only slightly concave, and showery if the concavity is deep.—BACON.

Grey. A grey sky in the morning presages fine weather.—FITZROY.

Clouds. If at sunrising the clouds are driven away, and retire, as it were, to the west, this denotes fair weather.—PLINY.

If at sunrise small reddish-looking clouds are seen low on the horizon, it must not always be considered to indicate rain. The probability of rain under these circumstances will depend on the character of the clouds and their height above the horizon. I have frequently observed that if they extend 10°, rain will follow before sunset; if 20° or 30°, rain will follow before 2 or 3 p.m.; but if still higher and near the zenith, rain will fall within three hours.—C. L. PRINCE.

Clouds like globes at sunrise announce clear, sharp weather.

[Sunrise.]

Above the rest, the sun who never lies,
Foretells the change of weather in the skies;
For if he rise unwilling to his race,
Clouds on his brow and spots upon his face, *Clouds.*
Or if through mists he shoot his sullen beams,
Frugal of light in loose and straggling streams,
Suspect a drizzling day and southern rain,
Fatal to fruits, and flocks, and promised grain.
VIRGIL'S "GEORGICS," BOOK I., LINE 438.

Clear, etc. A high dawn indicates wind. A low dawn indicates fair weather.

[*Note.*—A high dawn is when the first indications of daylight are seen over a bank of clouds; a low dawn is when the day breaks on or near the horizon, the first streaks of light being very low down.—FITZROY.]

Cloudy. Clouds collected near the sun at sunrise forebode a rough storm that same day; but if they are driven from the east and pass away to the west, it will be fine.—BACON.

If at sunrise the clouds about the sun disperse, some to the north and some to the south, though the sky round the sun itself is clear, it portends wind.—BACON.

If the sky at sunrise is cloudy and the clouds soon disperse, certain fine weather will follow.—SHEPHERD OF BANBURY.

Gloomy.

If Aurora, with half-open eyes,
And a pale, sickly cheek, salutes the skies
How shall the vine with tender leaves defend
Her teeming clusters when the storms descend?
VIRGIL.

Stormy. Storms are said to decrease at the rising or setting of the sun or moon.

Misty. A general mist before the sun rises near the full moon presages fair weather.—SHEPHERD OF BANBURY.

Pale. The sun pale and (as we call it) watery at its rising denotes rain; if it set pale, wind.—BACON.

Rays. If at sunrise the sun emits rays from the clouds, the middle of his disc being concealed therein, it indicates rain, especially if these rays break out downwards, so as to make the sun appear bearded. But if rays strike from the centre, or from different parts of the sun, whilst the outer circle of his disc is covered with clouds, there will be great storms both of wind and rain.—BACON.

Proverb. A morning sun, a wine-bred child, and a Latin-bred woman seldom end well.—G. HERBERT.

[Sunrise.] Sunny. Cloudy.

A glaring, sunny morning never comes to a good end.
FRENCH.

If at sunrise the clouds do not appear to surround the sun, but to press upon him from above, as if they were going to eclipse him, a wind will arise from the quarter on which the clouds incline. If this take place at noon, the wind will be accompanied by rain.—BACON.

Gaudy.

A gaudy morning bodes a wet afternoon.

Red morning.

Or if Aurora tinge with glowing red
The clouds that float round Phœbus' rising head.
Farmer, rejoice! for soon refreshing rains
Will fill the pools and quench the thirsty plains.
If ere his limbs he rear from ocean's bed

Dark clouds.

His foremost rays obscure and dark are spread
On th' horizon's edge, forewarned, take heed;
These signs the rain or blustering wind precede.
J. LAMB'S "ARATUS."

Red.

If the clouds at sunrise be red, there will be rain the following day.

In the winter season, a red sky at sunrise foreshows steady rain on the same day. The same sign in summer betokens occasional violent showers, wind in both cases generally accompanying.

A red morn, that ever yet betokened
Wreck to the seaman, tempest to the field,
Sorrow to shepherds, woe unto the birds,
Gust and foul flaws to herdmen and to herds.
SHAKESPEARE.

If red the sun begin his race,
Be sure the rain will fall apace.

Ruddy.

If the rays of the sun on rising are not yellow, but ruddy, it denotes rain rather than wind. The same likewise holds good of the setting.—BACON.

[SUNSET.]

But more than all the setting sun survey,
When down the steep of heaven he drives the day;
For oft we find him finishing his race,
With various colours erring on his face.
If fiery red his glowing globe descends,
High winds and furious tempests he portends;
But if his cheeks are swoln with livid blue,
He bodes wet weather by his watery hue;
If dusky spots are varied on his brow,
And streaked with red a troubled colour show,
That sullen mixture shall at once declare
Winds, rain, and storms, and elemental war.

*   *   *   *   *

But if with purple rays he brings the light,
And a pure heaven resigns to quiet night,
No rising winds or falling storms are nigh.—VIRGIL. [*Sunset.*]

Breeze. A breeze usually springs up before sunset; or if a gale is blowing, it generally subsides about that time.

Clear.

Sun set in a clear,
Easterly wind's near;
Sun set in a bank,
Westerly will not lack.
ST. ANDREWS, SCOTLAND.

Bright.

When the sun sets bright and clear,
An easterly wind you need not fear.

Red If the sun set with a very red eastern sky, expect wind; if red to the south-east, expect rain.

When Tottenham Wood is all on fire,
Then Tottenham Street is nought but mire.
MIDDLESEX.

If the body of the sun appear blood red at setting, it forebodes high winds for many days.—BACON.

Red west at sunset, not extending far up the sky, and having no thick bank of black clouds, will be followed by a fine day.

Colours of. When after sunset the western sky is of a whitish yellow, and this tint extends a great height, it is probable that it will rain during the night or the next day. Gaudy or unusual hues, with hard, definitely outlined clouds, foretell rain, and probably wind. If the sun before setting appears diffuse and of a brilliant white, it foretells storm. If it sets in a sky slightly purple, the atmosphere near the zenith being of a bright blue, we may rely on fine weather.

Rhyme.

If the sun in red should set,
The next day surely will be wet;
If the sun should set in grey,
The next will be a rainy day.

Golden.

The weary sun hath made a golden set,
And by the bright track of his fiery car
Gives token of a goodly day to-morrow.
SHAKESPEARE'S "RICHARD III."

When the sun sets of a golden yellow colour, with disc ill defined, and rays extending 4° or 6°, a strong wind and much vapour exist at a considerable elevation, and rain usually occurs within twenty-four hours.—C. L. PRINCE.

Yellow. A bright yellow sky at sunset presages wind; a pale yellow, wet.—FITZROY.

[Sunset.] Hazy. When the air is hazy, so that the solar light fades gradually, and looks white, rain will most certainly follow.

In summer time, when the sun at rising is obscured by a mist which disperses about three hours afterwards, expect hot and calm weather for two or three days.—C. L. PRINCE.

Pale.

If the sun goes pale to bed,
'Twill rain to-morrow, it is said.

When the sun appears of a light pale colour, or goes down into a bank of clouds, it indicates the approach or continuance of bad weather.

Sad. When the sun sets sadly, the morning will be angry.

ZUÑI INDIANS.

Cloudy. Black or dark clouds arising at sunset prognosticate rain,—on the same night, if they rise in the east opposite the sun; if close to the sun in the west, the next day, accompanied with wind.—BACON.

The sun setting behind a cloud forebodes rain the next day; but actual rain at sunset is rather a sign of wind. If the clouds appear as if they were drawn towards the sun, it denotes both wind and rain.—BACON.

When the sun sets in a bank,
A westerly wind we shall not lack.

The sun setting after a fine day behind a heavy bank of clouds, with a falling barometer, is generally indicative of rain or snow, according to the season, either in the night or next morning. In winter, if there has been frost, it is often followed by thaw. Sometimes there will be a rise of temperature only, no rain falling to any amount.—JENYNS.

Wet.

The sun sets weeping in the lowly west,
Witnessing storms to come, woe and unrest.

SHAKESPEARE'S "RICHARD II."

[SUNRISE AND SUNSET.]

Red.

The skie being red at evening,
Foreshewes a faire and cleare morning;
But if the morning riseth red,
Of wind and raine we shall be sped.—A. FLEMING.

Rose tints at sunset and grey dawn, a fine day to follow.

Sunrise full.

If Phœbus rising wide and broad appear,
And as he mounts contracts his ample sphere,
Propitious sign, no rain or tempest near.
Propitious, too, if after days of rain

Sunset pale.

With a pale face he seek the western main.

Cloudy.

When through the day the angry welkin lowers,
Hid is the sun, and drenched the earth with showers,
Catch if thou canst his last departing ray,
And gain prognostics of the following day.

[Sunrise and Sunset.]

Sunset with black cloud.

If by black cloud eclipsed his orb is found
Shooting his scattered rays at random round,
Send not the traveller from thy roof away—
To-morrow shines no brighter than to-day.

Sunset clear. *Crimson.*

If with clear face into his watery bed,
Curtained with crimson clouds around his head,
He sink, that night no rain or tempest fear;
And morrow's sun will shine serene and clear.

J. LAMB'S "ARATUS."

When it is evening, ye say, It will be fair weather: for the sky is red. And in the morning, It will be foul weather to-day: for the sky is red and lowring.—MATTHEW xvi. 2, 3.

Sunrise or Sunset.

If when the sun begin his daily race,
Or ere he sink in ocean's cool embrace,

Rays.

The rays that crown his head together bend,
And to one central point converging tend;

Clouds.

Or if by circling clouds he is opprest,
Hanging about him as a vapoury vest;

Little cloud.

Or if before him mount a little cloud,
Veiling his rising beams in murky shroud,—
By these forewarned within the house remain;
Charged is the air with stores of pelting rain.

J. LAMB'S "ARATUS."

Grey and Red.

An evening grey and a morning red
Will send the shepherd wet to bed.

Evening grey and morning red
Make the shepherd hang his head.

Evening red and morning grey,
Two sure signs of one fine day.

If the evening is red and the morning grey,
It is the sign of a bonnie day;
If the evening's grey and the morning red,
The lamb and the ewe will go wet to bed.—YARROW.

Sky red in the morning
Is a sailor's sure warning;
Sky red at night
Is the sailor's delight.

A red evening and a grey morning set the pilgrim a-walking.

ITALY.

An evening red and morning grey make the pilgrim sing.

FRANCE.

[Sunrise and Sunset.] Grey and Red.

Evening red and morning grey
Help the traveller on his way;
Evening grey and morning red
Bring down rain upon his head.

The evening red and the morning grey
Is the sign of a bright and cheery day;
The evening grey and the morning red,
Put on your hat, or you'll wet your head.—SCOTLAND.

Dull. If either on rising or setting the sun's rays appear shortened or contracted, and do not shine out bright, though there are no clouds, it denotes rain rather than wind.—BACON.

Lurid. If the sun on rising or setting cast a lurid red light on the sky as far as the zenith, it is a sure sign of storms and gales of wind.

Cloudy. When clouds are tinged on their *upper* edge of a pink or copper colour, and situated to the eastward at sunset, or to the westward at sunrise, expect wind and rain in about forty-eight hours—seldom much earlier.—C. L. PRINCE.

Sun.

Next mark the features of the God of Day;
Most certain signs to mortals they convey,
When fresh he breaks the portals of the east,
And when his wearied coursers sink to rest.
Sunrise bright. If bright he rise, from speck and tarnish clear,
Throughout the day no rain or tempest fear.
Sunset cloudless. If cloudless his full orb descend at night,
To-morrow's sun will rise and shine as bright.
Sunrise, dark cloud. But if returning to the eastern sky,
A hollow blackness on his centre lie;
Sunbeams north and south. Or north and south his lengthened beams extend,—
These signs a stormy wind or rain portend.
Without rays visible. Red. Observe if shorn of circling rays his head,
And o'er his face a veil of redness spread;
For * o'er the plains the God of Winds will sweep,
Lashing the troubled bosom of the deep.
Dark. If in a shroud of blackness he appear,
Forewarned, take heed—a drenching rain is near.
Black and red (purple). If black and red their tints together blend,
And to his face a murky purple lend,
Soon will the wolfish wind tempestuous howl,
And the big cloud along the welkin roll.
And weather foul expect, when thou canst trace
Halo solar. A baleful halo circling Phœbus' face
Of murky darkness, and approaching near:
Double halo. If of two circles, fouler weather fear.

* Qy. Far?

[Sunrise and Sunset.]

Mark when from eastern wave his rays emerge,
And ere he quench them in the western surge,
If near th' horizon ruddy clouds arise,
Mocking the solar orb in form and size:
If two such satellites the sun attend,
Soon will impetuous rain from heaven descend:
If one, and north, the northern wind prevails;
If one, and south, expect the southern gales.

J. LAMB'S "ARATUS."

Red clouds round.

Double round red clouds.

Mock suns.

Mock suns predict a more or less certain change of weather.
SCOTLAND.

Solar halo.

When the sun is in his house [halo], it will rain soon.
ZUNI INDIANS.

If there be a ring or halo around the sun in bad weather, expect fine weather soon.

A bright circle round the sun denotes a storm and cooler weather.

A white ring round the sun towards sunset portends a slight gale that same night; but if the ring be dark or tawny, there will be a high wind the next day.—BACON.

If there be a circle round the sun at rising, expect wind from the quarter where the circle first begins to break; but if the whole circle disperses evenly, there will be fine weather.
BACON.

If the sun or moon outshines the "brugh" (or halo), bad weather will not come.

The circle of the moon never filled a pond; the circle of the sun wets a shepherd.

The bigger the ring, the nearer the wet.

Dog * before,
You'll have no more;
Dog behind,
Soon you'll find.

Eclipse.

Eclipse weather is a popular term in the south of England for the weather following an eclipse of the sun or moon, and it is vulgarly esteemed tempestuous and not to be depended on by the husbandman.

The hurricane eclipse of the sun.—CAMPBELL.

Eclipses of the moon are generally attended by winds, eclipses of the sun by fair weather, but neither of them are often accompanied by rain.—BACON.

* Sun dog or halo.—SHETLAND and SCOTLAND generally.

MOON.

Signs.

Each sign observe—more sure when two agree;
Nor doubt th' event foretold by omens three.
Note well the events of the preceding year,
And with the rising and the setting stars compare.
But chiefly look to Cynthia's varying face;
There surest signs of coming weather trace.

Obscured.

Observe when twice four days she veils her light,
Nor cheers with silvery ray the dreary night.
Mark these prognostics through the circling year,
And wisely for the rain, the wind, the storm, prepare.

J. LAMB'S "ARATUS."

Halo.

A halo oft fair Cynthia's face surrounds,
With single, double, or with triple bounds:

Single.

If with one ring, and broken it appear,
Sailors, beware! the driving gale is near.

Unbroken.

Unbroken if it vanisheth away—
Serene the air, and smooth the tranquil sea.

Double.

The double halo boisterous weather brings,

Triple.

And furious tempests follow triple rings.
These signs from Cynthia's varying orb arise—
Forewarn the prudent, and direct the wise.

J. LAMB'S "ARATUS."

Halo.

Far burr, near rain.—NAUTICAL.

[*Note.*—The farther the "burr" or halo appears from the moon, the nearer at hand is the coming rain.]

Circle* near, water far;
Circle far, water near.—ITALY.

A far brugh, a near storm.—SCOTCH.

[Meaning, a *distant* halo round the moon, a storm near at hand.]

When round the moon there is a brugh [halo],
The weather will be cold and rough.—SCOTLAND.

When the wheel is far, the storm is n'ar;
When the wheel is near, the storm is far.

The moon with a circle brings water in her beak.

The moon, if in house be, cloud it will, rain soon will come.
ZUÑI INDIANS.

Haloes round the moon, a blood-red sunset, a red moon on her fourth rising, . . . prognostics of winds.—BACON.

The open side of the halo tells the quarter from which the wind or rain may be expected.

* Halo round moon.

[Moon.]
Haloes.

Circles round the moon always foretell wind from the side where they break, and a remarkable brilliancy in any part of the circle denotes wind from that quarter.—BACON.

Double or treble circles round the moon foreshadow rough and severe storms, and much more so if these circles are not pure and entire, but spotted and broken.—BACON.

A circle or halo round the moon signifies rain rather than wind, unless the moon stand erect within the ring, when both are portended.—BACON.

For I fear a hurricane;
Last night the moon had a golden ring,
And to-night no moon we see.
LONGFELLOW'S "WRECK OF THE HESPERUS."

Haloes predict a storm (rain and wind, or snow and wind) at no great distance, and the open side of the halo tells the quarter from which it may be expected.—SCOTLAND.

If three days old her face be bright and clear, *Moon three days old.*
No rain or stormy gale the sailors fear;
But if she rise with bright and blushing cheek, *Bright.*
The blustering winds the bending mast will shake.
If dull her face and blunt her horns appear *Dull.*
On the fourth day, a breeze or rain is near. *Fourth day.*
If on the third she move with horns direct, *Third day.*
Not pointing downward or to heaven erect, *Moon "on her back."*
The western wind expect; and drenching rain,
If on the fourth her horns direct remain. *Horns inclined.*
If to the earth her upper horn she bend,
Cold Boreas from the north his blast will send;
If upward she extend it to the sky,
Loud Notus with his blustering gale is nigh.
When the fourth day around her orb is sprea d
A circling ring of deep and murky red, *Halo.*
Soon from his cave the God of Storms will rise,
Dashing with foamy waves the lowering skies.
And when fair Cynthia her full orb displays, *Moon.*
Or when unveiled to sight are half her rays, *Half-moon.*
Then mark the various hues that paint her face, *Colours.*
And thus the fickle weather's changes trace.
If smile her pearly face benign and fair, *Bright.*
Calm and serene will breathe the balmy air;
If with deep blush her maiden cheek be red, *Rosy.*
Then boisterous wind the cautious sailors dread;
If sullen blackness hang upon her brow, *Black.*
From clouds as black will rainy torrents flow.
Not through the month their power these signs extend,
But all their influence with the quarter end.
J. LAMB'S "ARATUS."

[Moon.]
Moonlight. Moonlight nights have the hardest frosts.

Clear. Clear moon,
Frost soon.—SCOTLAND.

Large. The moon appearing larger at sunset, and not dim, but luminous, portends fair weather for several days.—BACON.

Red, dim, or pale. A dim or pale moon indicates rain; a red moon indicates wind.

Dim. When the moon has a white look, or when her outline is not very clear, rain or snow is looked for.—SCOTLAND.

Ruddy. If on her cheeks you see the maiden's blush,
The ruddy moon foreshows that winds will rush.—VIRGIL.

Red. The moon, her face if red be,
Of water speaks she.—ZUÑI INDIANS.

Pale or red. Pale moon doth rain,
Red moon doth blow,
White moon doth neither rain nor snow.
FROM THE LATIN PROVERB (CLARKE, 1639).

Watery. When the moon is darkest near the horizon, expect rain.

The moon, methinks, looks with a watery eye.
SHAKESPEARE'S "MIDSUMMER NIGHT'S DREAM."

Influence. The labourer who believes in the influence of the moon will not fill his granary.—HAUTE LOIRE.

Rhyme. The moon and the weather
May change together;
But change of the moon
Does not change the weather.
If we'd no moon at all,
And that may seem strange,
We still should have weather
That's subject to change.
"NOTES AND QUERIES," SEPTEMBER 23RD, 1882.

Great or small. Moon changed, keeps closet three days as a queen
Ere she in her prime will of any be seen:
If great she appeareth, it showereth out;
If small she appeareth, it signifies drought.—TUSSER.

Fog. A fog and a small moon
Bring an easterly wind soon.—CORNWALL.

Way to wane. The three days of the change of the moon from the way to the wane we get no rain.—UNITED STATES.

Changes. If the moon changes with the wind in the east, the weather during that moon will be foul.

Five changes of the moon in one calendar month indicate cooler weather.

When changes of the moon occur in the morning, expect rain. *[Moon.] Changes.*

Moon changing in morning indicates warm weather; in the evening, cold weather.

If the moon is rainy throughout, it will be clear at the change, and perhaps the rain will return a few days after.

If the moon change on a Sunday, there will be a flood before the month is out.—WORCESTERSHIRE.

A Saturday moon,
If it comes once in seven years, comes once too soon.

Saturday's moon and Sunday's prime
Ance is aneugh in seven years' time.—SCOTLAND.

Saturday's change and Sunday's full
Never brought good and never wull.—NORFOLK.

A Saturday's change and a Sunday's full moon
Once in seven years is once too soon.

A Saturday's change and a Sunday's full
Comes too soon whene'er it wull.—DORSET.

A few days after full or new moon, changes of weather are thought more probable than at any other time.—SCOTLAND.

Waning.

In the decay of the moon
A cloudy morning bodes a fair afternoon.

Sowe peason and beans in the wane of the moone;
Who soweth them sooner, he soweth too soone.
TUSSER.

Mr. E. J. Lowe found that a red moonrise was followed seven times out of eight by rain. There were, only eight observations. *Moonrise red.*

When the moon rises red and appears large, with clouds, expect rain in twelve hours.

If she rises red, it portends wind; if reddish or dark-coloured, rain; but neither of these portend anything beyond the full.—BACON.

If the full moon rise pale, expect rain. *Pale rise.*

When the moon runs low, expect warm weather. *Low.*

When the moon runs high, expect cool or cold weather. *High.*

If the moon be fair throughout and rain at the close, the fair weather will probably return on the fourth or fifth day. *Fair.*

If the moon is seen between the scud and broken clouds during a gale, it is expected to cuff away the bad weather. *Gale moon.*

A dry moon is far north and soon seen. *Dry.*

[Moon.] Dry. The farther the moon is to the south, the greater the drought; the farther west, the greater the flood, and the farther north-west, the greater the cold.

Pale.

> Therefore the moon, the governess of floods,
> Pale in her anger, washes all the air,
> That rheumatic diseases do abound.
>
> SHAKESPEARE'S "MIDSUMMER NIGHT'S DREAM."

Seen in day. When the moon is visible in the daytime, the days are relatively cool.

Frost. Frost occurring in the dark of the moon kills fruit buds and blossoms, but frost in the light of the moon will not.

Rain moon. Confucius, the Chinese philosopher, in one of his walks advised his disciples to provide themselves with umbrellas, since, though the sky was perfectly fair, it would soon rain. This happened, and the sage said it was because he had read a verse of the *She King* to the effect that, when the moon rises in the constellation *pe*, great rain may be expected.

CHAMBERS' MISCELLANY.

Moon new. If at her birth, or within the first few days, the lower horn of the moon appear obscure, dark, or any way discoloured, there will be foul and stormy weather before the full. If she be discoloured in the middle, it will be stormy about the full; but if the upper horn is thus affected, about the wane.

BACON.

If the new moon appear with the points of the crescent turned up, the month will be dry. If the points are turned down, it will be wet.

[*Note.*—About one-third of the sailors believe in the direct opposite of the above. The belief is explained as follows:—Firstly, if the crescent will hold water, the month will be dry; if not, it will be wet. Secondly, if the Indian hunter could hang his powder-horn on the crescent, he did so, and stayed at home, because he knew that the woods would be too dry to still hunt. If he could not hang his powder-horn upon the crescent, he put it on his shoulder and went hunting, because he knew that the woods would be wet, and that he could stalk game noiselessly.—MAJOR DUNWOODY, U.S.]

Stormy wet weather. If there be a change from continued stormy or wet to clear and dry weather at the time of a new or full moon, it will probably remain fine till the following quarter; and if it changes not then, or only for a short time, it usually lasts until the following new or full moon; and if it does not change then, or only for a very short time, it will probably remain fine and dry for four or five weeks.

[Moon.] Phases.

If the new moon, first quarter, full moon, or last quarter, occur between the following hours, the weather here stated is said to follow :—

In summer between—

12 and 2 a.m. Fair.
2 and 4 a.m. Cold and showers.
4 and 6 a.m. Rain.
6 and 8 a.m. Wind and rain.
8 and 10 a.m. Changeable.
10 and 12 a.m. Frequent showers.
12 and 2 p.m. Very rainy.
2 and 4 p.m. Changeable.
4 and 6 p.m. Fair.
6 and 8 p.m. Fair, if wind N.W.
8 and 10 p.m. Rainy, if wind S. or S.W.
10 and 12 p.m. Fair.

In winter between—

12 and 2 a.m. Frost, unless wind S.W.
2 and 4 a.m. Snow and stormy.
4 and 6 a.m. Rain.
6 and 8 a.m. Stormy.
8 and 10 a.m. Cold rain, if wind W.
10 and 12 a.m. Cold and high wind.
12 and 2 p.m. Snow and rain.
2 and 4 p.m. Fair and mild.
4 and 6 p.m. Fair.
6 and 8 p.m. Fair and frosty, if wind N.E or N.
8 and 10 p.m. Rain or snow, if wind S. or S.W.
10 and 12 p.m. Fair and frosty.

UNITED STATES.

Thaws.

As many days from the first new moon, so many times will it thaw during winter.

North and south.

If the new moon is far north, it will be cold for two weeks; but if far south, it will be warm.

New.

New moon far in north, in summer, cool weather, in winter, cold.

New moon far in the south indicates dry weather for a month.

Horns sharp or dull.

A new moon with sharp horns threatens windy weather.

When Luna first her scattered fear recalls,
If with blunt horns she holds the dusky air,
Seamen and swains predict abundant showers.

VIRGIL.

If one horn of the moon is sharp and pointed, the other being more blunt, it rather indicates wind; but if both are so, it denotes rain.—BACON.

Sharp horns do threaten windy weather.

OLD PLAY QUOTED BY SWAINSON.

In winter, when the moon's horns are sharp and well defined, frost is expected.—SCOTLAND.

[Moon.] New. If the points of a new moon are up, then, as a rule, no rain will fall that quarter of the moon; a dull, pale moon, dry, with halo, indicates poor crops. In the planting season no grain must be planted when halo is around the moon.

APACHE INDIANS.

Bright. A uniform brightness in the sky at the new moon, or the fourth rising, presages fair weather for many days. If the sky is uniformly overcast, it denotes rain. If irregularly overcast, wind from the quarter where it is overcast. But if it suddenly becomes overcast without cloud or fog, so as to dull the brightness of the stars, rough and serious storms are imminent.—BACON.

Erect. An erect moon is almost always threatening and unfavourable, but principally denotes wind. If, however, she appear with blunt or shortened horns, it is rather a sign of rain.—BACON.

Moon on her back. People speak of the new moon lying on her back or being ill-made as a prognostic of wet weather.

New moon on its back indicates wind; standing on its point indicates rain in summer and snow in winter.

DR. JOHN MENUAL.

The bonnie moon is on her back;
Mend your shoes and sort your thack [thatch].

If the moon is on its back in the third quarter, it is a sign of rain.

When the moon lies on her back,
Then the sou'-west wind will crack;
When she rises up and nods,
Then north-easters dry the sods.

REVIEWER IN "SYMONS' METEOROLOGICAL MAGAZINE," SEPTEMBER 1867.

When the new moon lies on her back,
She sucks the wet into her lap.—ELLESMERE.

It is sure to be a dry moon if it lies on its back, so that you can hang your hat on its horns.—WELSH BORDER.

When first the moon appears, if then she shrouds
Her silver crescent tipped with sable clouds,
Conclude she bodes a tempest on the main,
And brews for fields impetuous floods of rain;
Or if her face with fiery flushings glow,
Expect the rattling winds aloft to blow;
But four nights old (for that's the surest sign)
With sharpened horns, if glorious then she shine,
Next day, nor only that, but all the moon,
Till her revolving race be wholly run,
Are void of temptests both by land and sea.—VIRGIL.

[Moon.] Snowstorm.

If a snowstorm begins when the moon is young, it will cease at moonrise.

Misty.

If mists in the new moon, rain in the old;
If mists in the old moon, rain in the new.
SHEPHERD OF BANBURY.

Change of.

From the first, second, and third days of the new moon nothing is to be predicted; on the fourth there is some indication; but from the character of the fifth and sixth days the weather of the whole month may be predicted.
MARSHAL BURGAND'S MOTTO.

Fourth day.

If the new moon is not visible before the fourth day, the air will be unsettled for the whole month.—BACON.

If on her fourth day the moon is clear, with her horns sharp, not lying entirely flat, nor standing quite upright, but something between the two, there is a promise mostly of fair weather till the next new moon.—BACON.

The prime or fourth day after the change of the moon doth most commonly determine the force and direction of the wind.
PLINY.

The dispositions of the air are shown by the new moon, though still more on the fourth rising, as if her newness were then confirmed. But the full moon itself is a better prognostic than any of the days which succeed it.—BACON.

As is the fourth and fifth day's weather,
So's that lunation altogether.—FROM THE LATIN.

Fifth day.

From long observation, sailors suspect storms on the fifth day of the moon.—BACON.

The weather remains the same during the whole moon:—

I. [Eleven times out of twelve] as it is on the fifth day, if it continues unchanged over the sixth day.

II. [Nine times out of twelve] as it is on the fourth day, if the sixth day resembles the fourth.
FRENCH—"GUARDIAN," SEPTEMBER 2ND, 1868.

Sixth day.

If the weather on the sixth day is the same as that of the fourth day of the moon, the same weather will continue during the whole moon.—SPANISH.

[Said to be correct nine times out of twelve.]

Dark part visible.

Late, late yestreen I saw the new moon
With the old moon in her arms;
And I fear, I fear, my master dear,
We shall have a deadly storm.
BALLAD OF SIR PATRICK SPENS.

[Moon.] Old.

To see the old moon in the arms of the new one is reckoned a sign of fine weather, and so is the turning up of the horns of the new moon.—SUFFOLK.

[In this position it is supposed to retain the water which is imagined to be in it.—NOTE BY SWAINSON.]

To see the old moon in the arms of the new one is a sign of bad weather to come.

Full.

Two full moons in a calendar month bring on a flood.
BEDFORDSHIRE.

The full moon eats clouds.—NAUTICAL.

The moon grows fat on clouds.

[*Note.*—The two last proverbs have arisen from a supposed clearance of clouds which is said to take place when the full moon rises. Close observation has, however, proved this to be an illusion.]

The weather is generally clearer at the full than at the other ages of the moon; but in winter the frost then is somctimes more intense.—BACON.

Full moons, with regard to colours and haloes, have, perhaps, the same prognostics as the fourth risings; but the fulfilment is more immediate, and not so long deferred.—BACON.

Acosta observes that in Peru, which is a bery windy country, there is most wind at the full moon.—BACON.

[*Note.*—There is no special prevalence of wind in Peru that I ever experienced.—R. I.]

In Western Kansas it is said that when the moon is near full it never storms.

When there are two full moons in one month, there are sure to be large floods.

Near full moon, a misty sunrise
Bodes fair weather and cloudless skies.

If the full moon rise red, expect wind.

The full moon brings fine weather.

April moon.

If from April 25th to 28th the full moon come with serene nights and no wind (at which times the dew commonly falls in great plenty), the ancients, from long experience, held it certain that the crops of grain would suffer.

PLAUTUS' "EPHÉMÉRIDES," ETC., EDITION OF 1556.

Clear.

If the moon show a silver shield,
Be not afraid to reap your field;
But if she rises haloed round,
Soon we'll tread on deluged ground.

[Moon.] *Old.*

If there be a general mist before sunrise near the full of the moon, the weather will be fine for some days.

Threatening clouds.

Threatening clouds, without rain, in old moon, indicate drought.

Mist.

Auld moon mist
Ne'er died of thirst.

An old moon in a mist
Is worth gold in a kist [chest];
But a new moon's mist
Will ne'er lack thirst.

STARS.

The obscuring of the smaller stars in a clear night is a sign of rain.—WING'S "EPHEMERIS," 1649.

Huddling or mistiness.

When the stars begin to huddle,
The earth will soon become a puddle.

Wind.

Before the rising of a wind the lesser stars are not visible even on a clear night.—FROM PLINY, xviii. 80.

The stars twinkle; we cry "Wind."—MALTA.

Twinkling.

Excessive twinkling of stars indicates heavy dews, rain, and snow, or stormy weather in the near future.

When stars flicker in a dark background, rain or snow follows soon.

Sky full of stars.

When the sky seems very full of stars, expect rain, or, in winter, frost.

Arcturus.

The prudent mariner oft marks afar
The coming tempest by Boötes' star.
J. LAMB'S "ARATUS."

Superstitions respecting the stars near the moon.

A star dogging the moon (which is a rustic expression for a planet being for many nights persistently near the moon) foretells bad weather.

If a big star is dogging the moon, wild weather may be expected.

One star ahead of the moon, towing her, and another astern, chasing her, is a sure sign of a storm.—LANCASHIRE.

Stars in moon's halo.

Moon in a circle indicates storm, and number of stars in circle the number of days before storm.

Halo.

An entire circle round any planet or larger star forebodes rain; if the circle be broken, there will be wind from the quarter where it breaks.—BACON.

Pleiades.

If the Pleiades rise fine they set rainy, and if they rise wet they set fine.—SWAHILI PROVERB.

Pleiades and Hyades.

Rains and showers follow upon the rising of the Pleiades and Hyades, but without wind; storms upon the rising of Orion and Arcturus.—BACON.

[Stars.]
Nebula.
Phatne.

And when with deep-charged clouds the air's opprest,
Phatne, the spot that shines on Cancer's breast,
Attentive mark: if bright the spot appear,
Soon Phœbus smiles with face serene and clear,
Nor the returning rain and tempest fear.

J. LAMB'S "ARATUS."

Prœsepe.

If the cloud (nebula) called Prœsepe, or the manger, standing betwixt the Aselli,* do not appear when the air is serene and clear, it foreshows foul, cold, and winterly weather. If the northermost of these stars be hid, great winds from the south; but the other being hid, north-east winds.

WING'S "EPHEMERIS," 1649.

Stars dim.

When small stars, like those called Aselli, are not visible in any part of the sky, there will be great storms and rains within a few days; but if these stars are only obscured in places, and are bright elsewhere, they denote winds only, but sooner.

BACON.

Cancer.

Now mark where high upon the zodiac line
The stars of lustre-lacking Cancer shine.
Near to the constellation's southern bound

Nebula.

Phatne, a nebulous bright spot, is found.
On either side this cloud, nor distant far,
Glitters to north and south a little star.
Though not conspicuous, yet these two are famed—

Onoi or Aselli.

The Onoi by ancient sages named.
If when the sky around be bright and clear,
Sudden from sight the Phatne disappear,
And the two Onoi north and south are seen
Ready to meet—no obstacle between—
The welkin soon will blacken with the rain,
And torrents rush along the thirsty plain.
If black the Phatne, and the Onoi clear,
Sure sign again that drenching showers are near.
And if the northern star be lost to sight,
While still the southern glitters fair and bright,
Notus will blow. But if the southern fail,
And clear the northern, Boreas will prevail.
And as the skies above, the waves below
Signs of the rising wind and tempest show.

J. LAMB'S "ARATUS."

* Two stars in Cancer.

[Stars.] Fading.

When the bright gems that night's black vault adorn
But faintly shine—of half their radiance shorn—
And not by cloud obscured or dimmed to sight
By the fine silvery veil of Cynthia's light,
But of themselves appear to faint away,
They warning give of a tempestuous day.
J. LAMB'S "ARATUS."

Milky Way.

The edge of the Milky Way which is brightest indicates the direction from which an approaching storm will come.
UNITED STATES.

Planets' conjunctions.

Wind must be expected both before and after the conjunctions of all the other planets with one another, except the sun; but fair weather from their conjunctions with the sun.—BACON.

COMETS.

Comets are said to bring cold weather.

Wine.

Comets are said to improve the grape crop; and wine produced in years when comets appear is called "comet wine."
FRENCH.

Omens.

All comets evidence the approach of some calamity, such as drought, famine, war, floods, etc.—APACHE INDIANS.

No grateful sight to husbandmen appear
One or more comets, with their blazing hair—
Forerunners of a parched and barren year.
J. LAMB'S "ARATUS."

METEORS.

If many meteors in summer, expect thunder.

Many meteors presage much snow next winter.

Numerous.

If meteors shoot toward the north, expect a north wind next day. Many shooting stars on summer nights indicate hot weather; in winter, a thaw.

After an unusual fall of meteors, dry weather is expected.

Mark when athwart the ebon vault of night
The meteors shoot their flash of vivid light—
From that same quarter will the wind arise,
And in like manner rush along the skies.
If numerous and from various points they blaze,
Darting across each other's paths their rays,
From various points conflicting winds will sweep
In whirlwind fury o'er the troubled deep.
J. LAMB'S "ARATUS."

Numerous falling stars presage wind next day.—SCOTLAND.

Streams.

Professor Erman, of Berlin, ascribes the spell of cold usually felt about May 10th, and also about August 10th, November 13th, and between February 5th and 11th, to the meteor streams which the earth's orbit crosses at these times.

[Meteors.] Shooting stars, as they are termed, foretell immediate winds from the quarter whence they shoot. But if they shoot from different or contrary quarters, there will be great storms both of wind and rain.—BACON.

AURORA. If an aurora appear during warm weather, cold and cloudy weather is to follow.—SCOTLAND.

Bright. The aurora, when very bright, indicates approaching storm.

Storm. The first great aurora, after a long tract of fine weather in September or beginning of October, is followed on the second day, and not till the second day about one o'clock, on the east coast, and about eleven o'clock in Nithsdale, by a great storm ; the next day after the aurora is fine weather.
PROFESSOR CHRISTISON (SCOTLAND).

Change. The aurora borealis indicates approaching change.

St. Elmo's fire. The ball of fire, called Castor by the ancients, that appears at sea, if it be single, prognosticates a severe storm, which will be much more severe if the ball does not adhere to the mast, but rolls or dances about. But if there are two of them, and that, too, when the storm has increased, it is reckoned a good sign. But if there are three of them, the storm will become more fearful.—BACON, FROM PLINY, ii. 37.

Last night I saw St. Elmo's stars,
With their glimmering lanterns all at play,
On the tops of the masts and the tips of the spars,
And I knew we should have foul weather that day.
[Also called Cuerpo Santo, Corposant, and Pey's Aunt by the fishermen.]

Wind.

A mass of weather wisdom has accumulated respecting the wind. It is generally more of a descriptive than of a prophetic character, but will serve to indicate to the acute observer of nature the kind of weather to expect when ever so small a change takes place in the direction or force of the wind.

WIND. There is more sea to the south and more land to the north, which likewise has no slight influence upon the winds.
BACON.

Governing weather. Every wind has its weather.—BACON.

Lord Rutherford and Lord Cockburn were once rambling on the Pentland Hills, and they complained to an old shepherd whom they met of the keenness of the wind. He could find no fault with it; and on their asking him why he

Uses. approved of it, he replied, "Weel, it dries the yird [soil], it slockens [refreshes] the ewes, and it's God's wull."

No weather is ill,
If the wind be still.

[Wind.] Quiet.

Bringing weather.

Look not, like the Dutchman, to leeward for fine weather.

Swift.

Blow the wind never so fast,
It will fall at last.—T. FULLER.

Sudden gusts.

Sudden gusts never come in a clear sky, but only when it is cloudy and with rain.—BACON.

Strong.

Strong winds are more uniform and regular than light breezes.
FITZROY.

Wind, clouds, and waves.

When a steady breeze of wind has continued to blow for any length of time, with a clear sky, or small clouds high in the atmosphere, the waves are generally regular and smooth, gliding in the direction of the wind, particularly when there is no current. At such times, if a dense cloud is generated, and is low in the atmosphere when passing over the observer, the strength of the regular breeze is decreased, and the waves appear to be agitated by the cloud whilst it passes over them, their summits being more elevated and turbulent. But no sooner has the dense cloud passed the zenith of the observer, than the breeze resumes its former strength, and the waves glide along as smooth as before.
NICHOLSON'S JOURNAL.

Increasing.

If the wind increases during a rain, fair weather may be expected soon.

Rise and fall.

The smaller and lighter winds generally rise in the morning and fall at sunset.—BACON.

Day and night.

The winds of the daytime wrestle and fight
Longer and stronger than those of the night.

A capful of wind.

In Sir Walter Scott's novel of *The Pirate* there is a note about King Eric (also called Windy Cap), who could change the direction of the wind by merely turning his cap round upon his head. Old Scotch women are also mentioned who, for a consideration, would promise to bring the wind from any desired quarter; and in the same novel Norna of the Fitful Head professed to control the wind by merely waving her wand in the air.

Hodnet.

As soon as Hodnet sends the wind,
A rainy day will Drayton find.—SHROPSHIRE.

When the cock has his neb in Hodnet Hole, look out for rain. [This refers to the weathercock on Drayton Church, whence Hodnet lies south-west.]
GEORGINA JACKSON'S "SHROPSHIRE FOLK-LORE."

[Wind.] Habberley. A storm will go three miles out of its way to come by Habberley to Churton [Church Pulverbatch].

GEORGINA JACKSON'S "SHROPSHIRE FOLK-LORE."

Night.

Winds at night are always bright;
But winds in the morning, sailors take warning.

A wind generally sets from the sea to the land during the day, and from the land to the sea at the night, especially in hot climates.—J. F. DANIELS.

Storms. Wind storms usually subside about sunset; but if they do not, they will go on for another day.

Brisk. A brisk wind generally precedes rain.

Rain.

For raging winds blow up incessant showers;
And when the rage allays, the rain begins.

SHAKESPEARE'S "HENRY VI."

Ripple of. There is a peculiar rippling of the wind, or broken way of blowing, which is said always to prognosticate heavy rain within a few hours.—SCOTLAND.

Wind and rain.

When rain comes before wind,
Halyards, sheets, and braces mind;
But—
When wind comes before rain,
Soon you may make sail again.—FITZROY.

When the rain comes before the winds,
You may reef when it begins;
But when the wind comes before the rain,
You may hoist your topsails up again.

If the rain comes before the wind,
Lower your topsails and take them in;
If the wind comes before the rain,
Lower your topsails and hoist them again.

When the rain's before the wind,
Your topsail halyards you must mind;
But when the wind's before the rain,
You may hoist your topsails up again.

CAPTAIN NARES.

Showers generally allay the winds, especially if they be stormy; as, on the other hand, winds often keep off rain.

BACON.

Oft is there use of winds that loud
Are whistling o'er the plains;
And oft of heaven-descending rains,
Daughters of the stormy cloud.

CARY'S "PINDAR."

If rain falls before the wind commences, the wind will last longer than the rain. But if the wind blows first, and is afterwards laid by rain, it does not often rise again; and if it does, it is followed by fresh rain.—BACON.

[*Wind.*] *Rain.*

Much wind brings rain.—FRENCH.

Therefore the winds have sucked up from the sea
Contagious fogs, which, falling in the land,
Have every pelting river made so proud,
That they have overborne their continents.
SHAKESPEARE'S "MIDSUMMER NIGHT'S DREAM."

Changing.

If the wind shifts about for a few hours, as if it was trying the different points, and then commences to blow constantly from one quarter, that wind will last many days.—BACON.

Backing.

When the wind backs and the weather glass falls,
Then be on your guard against gales and squalls.

Winds that change against the sun
Are always sure to backward run.

Veering N.W.S.E.

When the wind veers against the sun,
Trust it not, for back 'twill run.

The veering of the wind *with* the sun, or, as sailors say, right handed, prognosticates drïer or better weather; the *backing* of the wind *against* the sun, or left handed shifting, indicates rain, or more wind, or both together.—FITZROY.

A veering wind, fair weather.
A backing wind, foul weather.

N.E.S.W.

It is a sign of continued fine weather when the wind changes during the day so as to follow the sun.

If wind follows sun's course, expect fair weather.

Dove's law.

Permanent winds turn the vane only in a direct sense or *with* the sun.—DOVÉ.

In northern hemisphere the wind changes from east to west by way of south, and the reverse (from east to west by way of north) in the southern hemisphere.—DOVÉ.

In a note by Mr. E. Poste, author of *The Skies and Weather Forecasts of Aratus*, a passage is quoted as showing an anticipation of Dové's law. Aratus writes of—

"Veering winds,
Unstable, baffling the predictor's skill."

Theophrastus had before penned the following sentence on the subject (I quote Mr. Poste's translation): "When winds are not arrested by other winds (this is a confession of some undefined perturbations), but cease of themselves, they are transformed into the adjacent winds, rotating from left to right, like the sun in his (diurnal) course."

[Wind.] *Dové's law.* Theophrastus has taken this from his master Aristotle, who says: "The cycle of the winds, when they cease of themselves (*i.e.*, without being disturbed by opposite winds), is a continuous transformation of wind from one quarter into a wind from the adjacent quarter, following the direction of the (diurnal) movement of the sun." So that we are indebted to Mr. Poste for pointing out that these philosophers knew of the law by which permanent winds in the northern hemisphere turn, as the sailors say, "with the sun." That this law should have been rediscovered by Dové so many centuries after is a tribute to the accuracy and intelligence of the ancient observers.

With sun. If the wind follow the motion of the sun—that is, if it move from east to south, from south to west, from west to north, from north to east—it does not generally go back; or if it does, it is only for a short time. But if it move contrary to the sun—that is, if it changes from east to north, from north to west, from west to south, from south to east—it generally returns to the former quarter, at least before it has completed the entire circle.—BACON.

At sunset. If in unsettled weather the wind veers from south-west to west or north-west at sunset, expect finer weather for a day or two.—FITZROY.

North to north-east. If the wind veers from north to north-east in winter, intense cold follows.—DOVÉ.

Cyclones *Ballot's law.* Cyclones in northern hemisphere veer generally from east to west by way of north, or against the sun's course. In the southern hemisphere the reverse.—BUYS BALLOT.

[To remember this, think of the words NOT and SAME, meaning that winds change *not* in the N. hemisphere and *same* in S. hemisphere as the sun; or if preferred, one may consider a watch dial as laid horizontally, and the cyclonic wind will change not in northern and same in southern hemisphere as the movement of the hands. The N. and S. call to mind the rule as applying to the N. or S. hemisphere.—R. I.]

A cyclone in the torrid zone is always preceded by a fall in the barometer, and generally also by a greasy halo round the sun or moon, by rolled and tufted clouds with lurid streaks of light and unusual colours, and by a heavy bank of cloud clinging to the horizon, and often darting out threads of pale lightning.

Vortex. To find out where the centre or vortex of a cyclone is situated look to the wind's eye; set its bearing by the compass, and the eighth point (at 90°) to the RIGHT thereof will in the *northern* hemisphere be the bearing of the storm centre. The eighth point (or 90°) to the LEFT will be the same in the *southern* hemisphere.

[Wind.] Cyclones.

This wind is said to go "withershins," or contrary to the course of the sun.—SWAINSON.

Withershins.

I have several times, in calm weather, seen a cloud generate and diffuse a breeze on the surface of the sea, which spread in different directions from the place of descent. A remarkable instance of this occurred in Malacca Strait during a calm day, when a fleet was in company. A breeze commenced suddenly from a dense cloud; its centre of action seemed to be in the middle of the fleet, which was much scattered. This breeze spread in every direction from a centre, and produced a singular appearance in the fleet; for every ship hauled close to the wind as the breeze reached her, and when it became general exhibited to view the different ships sailing completely round a circle, although all hauled close to the wind.—NICHOLSON'S JOURNAL.

Cloud.

Cruel storms do not blow in a right course.—STRABO.

Storms cyclonic.

Cyclones generally move as a whole to the westward, curving to the northward, in northern latitudes; and to the westward, curving to the south, in southern latitudes.

Movement.

Cyclones are most violent near their centres.

The forceful whirlwind veers around.
POTTER'S "EURIPIDES."

June—too soon;
July—stand by;
August—look out you must;
September—remember;
October—all over.—CAPTAIN NARES.

Hurricanes in West Indies.

Squalls are considered as a favourable sign in tempests and hurricanes, as shortly preceding their discontinuance. They are accessions of new air to the prevailing wind or storm, and partly from a new direction, and are generally accompanied by arched clouds, or thunderstorms, and by rain.
FITZROY.

Squalls.

A storm moderates, to storm again.

Storm.

Untimely storms make men expect a dearth.
SHAKESPEARE'S "RICHARD III."

Storms unseasonable.

As humorous as winter, and as sudden
As flaws congealed in the spring of day.
SHAKESPEARE'S "HENRY IV."

In morning.

[Wind.] Sudden. The sudden storm lasts not three hours.

The sharper the blast
The sooner 'tis past.—CHARLES WESLEY.

North to south, and vice versâ. The wind usually turns from north to south with a quiet wind without rain, but returns to the north with a strong wind and rain. The strongest winds are when it turns from south to north by west.—FITZROY.

North-east to east. When the wind turns from north-east to east, and continues two days without rain, and does not turn south the third day, nor rain the third day, it is likely to continue north-east for eight or nine days, all fair, and then to come to the south again.—FITZROY.

South to north. If the wind shifts from south to north through west, there will be, in winter, snow; in spring, sleet; in summer, thunder-storms, after which the air becomes colder.—DOVÉ.

Changing. The wind goeth toward the south, and turneth about unto the north; it whirleth about continually, and the wind returneth again according to his circuits.—ECCLESIASTES i. 6.

Shifting during drought. In Texas and the south-west, when the wind shifts during a drought, expect rain.

Unsteadiness. Unsteadiness of wind shows changing weather.

A frequent change of wind, with agitation in the clouds, denotes a storm.

The often changing of the wind doth many times show stormy weather.—WING, 1649.

Sudden changes.

And more inconstant than the wind, who woos
Even now the frozen bosom of the north;
And being angered, puffs away from thence,
Turning his face to the dew-dropping south.
SHAKESPEARE'S "ROMEO AND JULIET."

Night changes. Winds changing from foul to fair during the night are not permanent.

Sudden shift. The wind having held long and extremely sharp in one point, and at last suddenly shifting, brings a relaxation, if not a thorough thaw.—POINTER.

Air currents. Currents of air frequently change their course, first in the higher regions, and are afterwards continued in other directions on the earth's surface, whence we can often foresee a change of wind by observing the clouds. Both the strength of a coming gale, and the point from which it will blow, may usually be determined by noticing the velocity and direction of the clouds floating along in the upper currents. *[Wind.]*

Calm. Always a calm before a storm.

After a storm comes a calm.

Changes.

Lang foul,
Lang fair.
BUCHANAN'S ALMANACK (SCOTLAND).

Various currents. In noticing the wind, regard must be had to whether there are one or more currents in the atmosphere: in the former case, the barometer is generally steady and the weather fair; in the latter, the mercury fluctuates and the weather is unsettled.—JENYNS.

Rolling. To discover the rolling cylinders of air, the vane of a weathercock might be so suspended as to dip or rise vertically, as well as to have its horizontal rotation.—E. DARWIN.

Tendency. Between the tropics winds and currents tend westward.

In middle latitudes winds and currents tend eastward.

In high latitudes winds and currents tend from the poles towards the equator.

Blight. It is certain that there are some blasts which leave behind them on plants manifest traces of burning and scorching. But the sirocco, which is an invisible lightning and a burning air without flame, is referred to the inquiry on lightning.
BACON.

Harmattan fog. A furious, scorching African wind, which is attended with a dense fog or haze.—E. DARWIN.

Mountains. Wherever there are high mountains covered with snow, periodical winds blow from that quarter at the time of the melting of the snows.—BACON.

Periodical. It has been remarked that periodical winds do not blow at night, but get up the third hour after sunrise.—BACON.

Light and heavy.

Light winds point to pressure low,
But gales around the same do blow.
ALEXANDER RINGWOOD.

Planets. Greater winds are observed to blow about the time of the conjunction of planets.—BACON.

Heat. If the wind be hushed with sudden heat, expect heavy rain.

The heat of the sun on its increase is more disposed to generate winds; on its decrease, to generate rain.—BACON.

Whispering. The whispering grove tells of a storm to come.

High. A high wind prevents frost.

[Wind.] Barley harvest. It is always windy in barley harvest; it blows off the heads for the poor.

Storm. If the wind is from the north-west or south-west, the storm will be short; if from the north-east, it will be a hard one; if from the north-west, a cold one; and if from the south-west, a warm one. After it has been raining some time, a blue sky in the south-east indicates that there will be fair weather soon.

Various.

North winds send hail, south winds bring rain,
East winds we bewail, west winds blow amain;
North-east is too cold, south-east not too warm,
North-west is too bold, south-west doth no harm.
The north is a noyer to grass of all suites,
The east a destroyer to herb and all fruits;
The south, with his showers, refresheth the corn;
The west to all flowers may not be forborne.
The west, as a father, all goodness doth bring;
The east, a forbearer, no manner of thing;
The south, as unkind, draweth sickness too near;
The north, as a friend, maketh all again clear.

TUSSER.

North. Wind from the north, cold and snow.

North-west. Wind from the western river of the north land, snow.

West. Wind from the world of waters, clouds.

South-west. Wind from the southern river of the world of waters, rain.

South. Wind from the land of the beautiful red, lovely odours and rain.

South-east. Wind from the wooded cañons, rain and moist clouds.

South. Wind from the land of day, it is the breath of health, and brings the days of long life.

North-east. Wind from the lands of cold bring the rain before which flees the harvest.

[The last eight are Indian proverbs, U.S.]

Direction of.

When the wind is in the north,
Hail comes forth.
When the wind is in the wast,
Look for a weet blast.
When the wind is in the soud,
The weather will be gude.
When the wind is in the east,
Cold and snaw come neist.—SCOTCH.

[Wind.] Direction of.

Wind east or west
Is a sign of a blast;
Wind north or south
Is a sign of a drought.

North wind cold,
East wind dry,
South wind warm and often wet,
West wind generally rainy.—BACON.

Satire.

The south wind always brings wet weather,
The north wind wet and cold together;
The west wind always brings us rain,
The east wind blows it back again;
If the sun in red should set,
The next day surely will be wet;
If the sun should set in grey,
The next will be a rainy day.

SATIRE ON THE HUMID CLIMATE OF THE BRITISH ISLES.

Drought and blast.

North and south, the sign o' drouth;
East and west, the sign of blast.

North, bad for fishers.

When the wind is in the north,
The skilful fisher goes not forth.

West, good for fishers.

Fishermen in anger froth
When the wind is in the north;
For fish bite the best
When the wind is in the west.

North.

When the wind's in the north,
You mustn't go forth.—DENHAM.

Fair.

A northern air
Brings weather fair.

Fair weather cometh out of the north.—JOB xxxvii. 22.

The gold [of the sky] cometh out of the north.
THE SAME, SHARPE'S TRANSLATION.

Cold.

And cold out of the north.—JOB xxxvii. 9.

To run upon the sharp wind of the north,
To do me business in the veins o' the earth
When it is backed with frost.
SHAKESPEARE'S "TEMPEST."

Rainy.

The north wind bringeth forth rain.
PROVERBS xxv. 23, SHARPE'S TRANSLATION.

Whirlwind.

A whirlwind came out of the north.—EZEKIEL i. 4.

Night.

The north wind, if it should rise by night (which is unusual) hardly ever lasts beyond three days.—BACON.

[Wind.] North. In large pastures shepherds should take care to drive their flocks to the north side, so that they may feed opposite to the south.—PLINY.

Grafting. Take care not to sow in a north wind, or to graft and inoculate when the wind is in the south.—PLINY.

Evils from north. All bad things come out of the north. A bleak, bad wind, and a biting frost, and a scolding wife come out of the north.

Channel. A north wind is a broom for the Channel.—CORNWALL.

First. Whenever the wind first blows from the north, after having been for some days in another direction, a fine day or two is almost sure to follow.

Snow.

The north wind doth blow,
And we shall have snow.—DENHAM.

Sterile. A north wind has no corn.—SPANISH.

Thunder. With a north wind it seldom thunders.

Cream. Cream makes most freely with a north wind.

New Moon. A new moon with north wind will hold until the full.

Changing. If there be within four, five, or six days two or three changes of wind from the north, through east without much rain and wind, and thence again through the west to the north with rain or wind, expect continued showery weather.

The north winds cease commonly after blowing an odd number of days—three, five, seven, or nine.—PLINY.

North-east. That the wind Cæcias [north-east] attracts clouds passed into a proverb among the Greeks.

ARISTOTLE'S "PROBLEMS," § DE VENTIS, 55.

If the wind is north-east three days without rain,
Eight days will pass before south wind again.

FITZROY.

Winds from the lands of cold bring fruit of ice. Wind from the right hand of the west is the breath of the god of sand clouds.—INDIAN PROVERBS, U.S.

North-west. Do business with men when the wind is in the north-west.

YORKSHIRE.

[*Note.*—This, bringing the finest weather, is said to improve men's tempers.]

Frost will probably occur when the temperature is 40° and the wind north-west.—UNITED STATES.

N.W. and S.W. A nor'-wester is not long in debt to a sou'-wester.

[Wind.] *N.W. or W., changing to S.W. or S., N.E. or E., to S.E. or S.*

If there be a change of wind from the north-west or west to the south-west or south, or else from the north-east or east to the south-east or south, expect fair weather.

UNITED STATES.

North-west and north-east.

When the wind is in the north-west,
The weather is at its best;
But if the rain comes out of the east,
'Twill rain twenty-four hours at least.

North-west wind brings a short storm; a north-east wind brings a long storm.

North-west, changing to south.

If the north-west or north winds blow with rain or snow during three or four days in the winter, and then the wind passes to the south through the west, expect continued rain.

N.W. to N.E. and S. and N.E.

If a north-west wind shifts to north-east, remaining there two or three days without rain, and then shifts to the south, and then back to the north-east, with very little rain, fair weather may be expected during the following month.

OBSERVER AT CAPE MENDOCINO.

North-west.

In summer, if the wind changes to the north-west, expect cooler weather.

North-west wind brings only rain showers.—UNITED STATES.

An honest man and a north-west wind generally go to sleep together.

[*Note.*—The north-west wind is said to abate at sunset.]

North-west and south-east.

If two currents of wind, as shown by the motions of the clouds, blow north-west and south-east respectively, and the south-east current be highest, foul weather will follow; but if the north-west current be uppermost, then fair, clear weather may be expected.

East.

When the wind is in the east,
It is neither good for man nor beast.

Dry.

The east wind dried up her fruit.—EZEKIEL xix. 12.

Their faces shall sup up as the east wind.—HABAKKUK i. 9.

An east wind shall come, the wind of the Lord shall come up from the wilderness, and his spring shall become dry, and his fountain shall be dried up.—HOSEA xiii. 15.

When the east wind toucheth it, it shall wither.

EZEKIEL xvii. 10.

And, behold, seven thin ears, and blasted with the east wind, came up.—GENESIS xli. 6.

Locusts.

The east wind brought the locusts.—EXODUS x. 13.

[Wind.] East. Dry.

A dry east wind raises the spring.—CORNWALL.

Easterly gales without rain during the spring equinox foretell a dry summer.—SCOTLAND.

Clear.

Everything looks large in the east wind.—SCOTLAND.

[*Note.*—There are many local sayings in Scotland referring to the unusually clear appearance of certain mountains during an east wind. It is said to indicate approaching rain.]

Cold.

When the hoar-frost is first accompanied by east wind, it indicates that the cold will continue a long time.

With rain.

When the rain is from the east,
It is for four-and-twenty hours at least.

An easterly wind's rain
Makes fools fain.

The heaviest rains begin with an easterly wind, which gradually veers round to south and west, or a little north-west, when the rain usually ceases.

Stormy.

God prepared a vehement east wind.—JONAH iv. 8.

The east wind hath broken thee in the midst of the seas.
EZEKIEL xxvii. 26.

Thou breakest the ships of Tarshish with an east wind.
PSALM xlviii. 7.

Thunder.

If an east wind blows against a dark, heavy sky from the north-west, the wind decreasing in force as the clouds approach, expect thunder and lightning.

East-north-east.

There arose against it a tempestuous wind, called Euroclydon.
ACTS xxvii. 14.

A tempestuous wind, called Euroclydon (or east-north-east).
THE SAME, SHARPE'S TRANSLATION.

East and north.

The east and north winds, when they have once begun, are more continuous; the south and west winds are more variable.—BACON.

East and west.

Wet weather with an east wind continues longer than with a west, and generally lasts a whole day.—BACON.

In an east wind all visible things appear larger; in a west wind all sounds are more audible and travel farther.
ARISTOTLE'S "PROBLEMS," § DE VENTIS, 55.

When the wind is in the east,
The fisher likes it least;
When the wind is in the west,
The fisher likes it best.

When the smoke goes west,
Gude weather is past;
When the smoke goes east,
Gude weather comes neist.—SCOTCH.

How thy garments are warm, when He quieteth the earth by the south wind.—JOB xxxvii. 17.

[Wind.] South (warm).

As whirlwinds in the south.—ISAIAH xxi. 1.

Tempestuous.

And shall go with whirlwinds of the south.
ZACHARIAH ix. 14.

Out of the south cometh the whirlwind.—JOB xxxvii. 9.

When ye see the south wind blow, ye say, There will be heat; and it cometh to pass.—LUKE xii. 55.

Hot.

Like foggy south, puffing with wind and rain.
SHAKESPEARE'S "AS YOU LIKE IT," ACT IV.

Foggy.

The weather usually clears at noon when a southerly wind is blowing.—NAUTICAL.

Noon.

When tempests of commotion like the south,
Born with black vapour, doth begin to melt,
And drop upon our bare, unarmèd heads.
SHAKESPEARE'S "HENRY IV."

Wet.

If the wind continue any considerable time in the south, it is an infallible sign of rain.—WING, 1649.

Continued.

If there be dry weather with a light south wind for five or six days, it having previously blown strongly from the same direction, expect fine weather.—TEXAS.

Light.

Brisk winds from the south for several days in Texas are generally followed by a "norther."

Brisk.

A southerly wind with a fog
Brings an easterly wind in snog [with certainty].
CORNWALL.

Foggy.

An out [southerly] wind and a fog
Bring an east wind home snug.—CORNWALL.

A southerly wind and a cloudy sky
Proclaim it a hunting morning.

Misty.

As when the south wind o'er the mountain tops
Spreads a thick veil of mist, the shepherd's bane,
But friendlier to the thief than shades of night.
HOMER'S "ILIAD."

In a south wind the sea appears more blue and clear, in a north wind blacker and darker.—ARISTOTLE.

Clear.

After frosts and long snows the south is almost the only wind that blows.—BACON, FROM ARISTOTLE'S "PROBLEMS," § DE VENTIS, 3.

In winter.

[Wind.] South. Rising and falling. When the south wind either rises or falls, there is generally a change of weather, from fair to cloudy, or from hot to cold, or *vice versâ.* But the north wind often both rises and falls without any change in the weather.—BACON.

Gentle. The south wind, when gentle, is not a great collector of clouds; but it is often clear, especially if it be of short continuance. But if it lasts or becomes violent, it makes the sky become cloudy and brings on rain, which comes on rather when the wind ceases or begins to die away, than when it commences or is at its height.—BACON.

Night. The south wind rises oftener and blows stronger by night than by day, especially in winter.—BACON, FROM ARISTOTLE.

Soothing.

The south wind warms the aged.

The south wind is the father of the poor.—RAGUSA.

When the wind's in the soud,
The weather will be fresh and gude;
When the wind's in the east,
Cauld and snaw come neist.

Damp.

And with the southern clouds contend in tears.
SHAKESPEARE'S "HENRY VI."

Rainy.

When the wind's in the south,
The rain's in its mouth.

The rain comes scouth [plentifully]
When the rain is in the south.—SCOTCH.

A southerly wind with showers of rain
Will bring the wind from west again.

Good for fishers.

When the wind is in the south,
It blows the bait in the fishes' mouth.

Fair. Fair weather for a week with a southern wind is likely to produce a great drought, if there has been much rain out of the south before.—FITZROY.

Whistling in leaves.

The southern wind
Doth play the trumpet to his purposes,
And by his hollow whistling in the leaves
Foretells a tempest and a blustering day.
SHAKESPEARE'S "HENRY IV."

North and south. If a south wind begin to blow for two or three days, a north wind will sometimes rise directly afterwards. But if there has been a north wind for as many days, the wind will blow for a short time from the east before it comes from the south.
BACON, FROM PLINY, ii. 48.

Towards the end of the year and the commencement of winter, if the south wind blow first and be succeeded by the north, it will be a severe winter (Arist., *Prob.*, xxvi. 49). But if the north wind blow at the commencement of winter, and be succeeded by the south, the winter will be mild and warm.—BACON.

[Wind.] North and south.

Rain with a south-east wind is expected to last for some time.
SCOTLAND.

South-east.

A south-west blow on ye,
And blister ye all over.
SHAKESPEARE'S "TEMPEST."

South-west (unwholesome).

Three south-westers, then one heavy rain.

Rainy.

In Southern Indiana a south-west wind is said to bring rain in thirty-six hours.

In fall and winter, if the wind holds a day or more in the south-west, a severe storm is coming; in summer the same may be said of a north-east wind.

Stormy.

The third day of south-west wind will be a gale, and wind will veer to north-west between 1 and 2 a.m. (in winter) with increasing force.—FISHERMEN OF NORTH CAROLINA.

Third day.

the wind shifts around to the south and south-west, expect warm weather.

Warm.

If the wind is south-west at Martinmas,
It keeps there till after Christmas.

Autumn.

after a stiff breeze there ensue a dead calm and drizzling rain, with a fall in the barometer, expect a gale from south-west.

Gale.

When the wind is in the west,
The weather is always best.

West.

The west wind is a gentleman, and goes to bed [*i.e.*, drops in the evening].

Wind west,
Rain's nest.—DEVONSHIRE.

Wet.

A western wind carrieth water in his hand.

A west wind, north about,
Never hangs lang out.—SCOTLAND.

Not permanent.

When the wind is on the west side of the compass, changes of barometer *accompany* changes of weather; but with the wind on the east side, the indications of the barometer *precede* the change.—G. F. CHAMBERS.

Rule.

[Wind.] West and east. The west wind is the attendant of the afternoon, for it blows more frequently than the east wind when the sun is declining.
BACON.

Calm. A dead calm often precedes a violent gale, and sometimes the calmest and clearest mornings in certain seasons are followed by a blowing, showery day. Calms are forerunners of the hurricanes of the West Indies and other tropical climes.

Clouds.

Clouds next come under notice, and it will be seen that much is to be gleaned by observing their forms and appearances. By Fitzroy, Howard, and others these masses of vapour have been marshalled in the order of their formation, so that the most casual obsetver may soon judge of the age of a cloud, whether seen in its early stage of light, misty stratus, or in the form of a dark, threatening nimbus, ripe for rain, and spreading like a vampire's wing over the landscape.

Although the names given by Howard to the different clouds have been here adopted, and the same general arrangement maintained, yet the familiar names given to these masses of vapour by sailors and others, such as Mackerel Sky, Mares' Tails, Wool Bags, Packet Boys, etc., have not been omitted. Clouds should of course be observed with a proper allowance for the force and direction of the wind at the time. With a swift upper current of air a clear sky sometimes becomes obscured in a few minutes, whilst in calmer weather changes in the appearance of the sky are slow to occur, and can be reckoned on with more safety.

In the frontispiece I have depicted such forms of clouds as are mentioned in this book, as well as some intermediate forms with the names of which I will not here trouble the reader. The clouds are arranged in the order of their height, so that the rough rule of "The higher the cloud, the finer the weather" may be more readily understood. The heights are those mentioned by Dr. Carl Lang and Dr. Fritz Erk in their recent report of 1891.*

CLOUDS.

And now the mists from earth are clouds in heaven,
Clouds slowly castellating in a calm
Sublimer than a storm, while brighter breathes
O'er the whole firmament the breadth of blue,
Because of that excessive purity
Of all those hanging snow-white palaces:
A gentle contrast, but with power divine.—WILSON.

Form. While any of the clouds, except the nimbus, retain their primitive forms, no rain can take place; and it is by observing

* *Deutsche Meteorigisches Jahrbuch* (Munich: 1891).

the changes and transitions of cloud form that weather may be predicted.—HOWARD. [*Clouds.*]

The higher the clouds, the finer the weather. *High.*

When on clear days isolated clouds drive over the zenith from the rain-wind side, storm and rain follow within twenty-four hours.—UNITED STATES. *Isolated.*

After clouds calm weather.—T. FULLER. *Calm.*

Clouds that the sun builds up darken him. *Dark.*

It it will not rain much so long as the sky is clear before the wind; but when clouds fall in against the wind, rain will soon follow. *With wind.*

When clouds break before the wind, leaving a clear sky, fine weather will follow.

After fine, clear weather the first signs in the sky of a coming change are usually light streaks, curls, wisps, or mottled patches of white distant clouds, which increase and are followed by an overcasting of murky vapour that grows into cloudiness. The appearance more or less oily or watery as wind or rain may prevail is an infallible sign. Usually the higher and more distant such clouds seem to be, the more gradual but general the coming change of weather will prove. *Indications of.*

FITZROY.

Now clouds combine, and spread o'er all the sky, *Growth of.*
When little rugged parts ascend on high,
Which may be twined, though by a feeble tie;
These make small clouds, which, driven on by wind,
To other like and little clouds are joined,
And these increase by more: at last they form
Thick, heavy clouds; and thence proceeds a storm.

CREECH'S "LUCRETIUS."

When clouds, after rain, disperse during the night, the weather will not remain clear. *Dispersing.*

Can any understand the spreadings of the clouds? *Spreading.*

JOB xxxvi. 29.

Dost thou know the balancing of the clouds?—JOB xxxvii. 16. *Balancing.*

Bleak is the morn when blows the north from high; *Dawn.*
Oft when the dawnlight paints the starry sky
A misty cloud suspended hovers o'er
Heaven's blessèd earth with fertilising store,
Drained from the living streams: aloft in air
The whirling winds the buoyant vapour bear,
Resolved at eve in rain or gusty cold,
As by the north the troubled rack is rolled.

ELTON'S TRANSLATION OF HESIOD'S WORKS.

[Clouds.] Morning.

Cloudy mornings turn to clear evenings.

When the clouds of the morn to the west fly away,
You may conclude on a settled, fair day.

Evening.

At sunset with a cloud so black,
A westerly wind you shall not lack.

Many small clouds at north-west in the evening show that rain is gathering, and will suddenly fall.—POINTER.

Storm cloud. When a heavy cloud comes up in the south-west, and seems to settle back again, look out for a storm.

Accumulating. If the sky, from being clear, becomes fretted or spotted all over with bunches of clouds, rain will soon fall.

SHEPHERD OF BANBURY.

Low.

If on the ocean's bosom clouds appear,
While the blue vault above is bright and clear,
These signs by shepherds and by sailors seen,
Give pleasing hope of days and nights serene.

J. LAMB'S "ARATUS."

Increasing. If clouds increase visibly, and the clear sky become less, it is a sign of rain.

Collecting and driving. If the clouds appear to drive fast when there is no wind, expect wind from that quarter from which they are driven. But if they gather and collect together, on the sun's approach to that part, they will begin to disperse; and then if they disperse towards the north, it prognosticates wind; if towards the south, rain.—BACON.

Driving.

When the carry [current of clouds] gaes west,
Gude weather is past;
When the carry gaes east,
Gude weather comes neist.

Clouds that are carried with a tempest, to whom the mist of darkness is reserved for ever.—2 PETER ii. 17.

From west. When ye see a cloud rise out of the west, straightway ye say, There cometh a shower; and so it is.—LUKE xii. 54.

Clearing. If the sky clears, and the clouds commence to break in the quarter opposite the wind, it will be fine; but if it clear up to windward, it indicates nothing, and leaves the weather uncertain.—BACON.

With mock suns. If clouds shall have shut in the sun, the less light there is left, and the smaller the sun's orb appears, the more severe will the storm prove. But if the disc of the sun appear double or treble, as if there were two or three suns, the storm will be much more violent, and will last many days.

BACON.

If the upper current of clouds comes from the north-west in the morning, a fine day will ensue. [Clouds.] North-west.

If in the north-west before daylight end there appear a company of small black clouds like flocks of sheep, it is a sure and certain sign of rain.—WING, 1649.

If a layer of thin clouds drive up from the north-west, and under other clouds moving more to the south, expect fine weather.—UNITED STATES.

Clouds in the east, obscuring the sun, indicate fair weather. *East.*

If clouds drive up high from the south, expect a thaw. *South.*

Small scattering clouds flying high in the south-west foreshow whirlwinds.—HOWARD. *South-west.*

A sky covered with clouds need not cause apprehension, if the latter are high, and of no great density, and the air is still, the barometer at the same time being high. Rain falling under such circumstances is generally light, or of not long continuance.—JENYNS. *High.*

If high, dark clouds are seen in spring, winter, or fall, expect cold weather.

Dark, heavy clouds, carried rapidly along near the earth, are a sign of great disturbance in the atmosphere from conflicting currents. At such times the weather is never settled, and rain extremely probable.—JENYNS. *Dark.*

If the clouds, as they come forward, seem to diverge from a point in the horizon, a wind may be expected from that quarter, or the opposite.—THOMAS BEST. *Diverging.*

The apparent permanency and stationary aspect of a cloud is often an optical deception, arising from the solution of vapour on one side of a given point, while it is precipitated on the other.—J. F. DANIELS. *Apparently stationary.*

Against heavy rain every cloud rises bigger than the preceding, and all are in a growing state.—G. ADAMS. *Rising.*

Clouds floating low, and casting shadows on the ground, are usually followed by rain.—UNITED STATES. *Low.*

High upper clouds, crossing the sun, moon, or stars in a direction different from that of the lower clouds, or the wind then felt below, foretell a change of wind toward their direction.
FITZROY. *Motions of.*

When the generality of the clouds rack or drive with the wind (though there are many in little fleeces, or long strakes lying higher, and appearing not to move), the wind is flagging, and will quickly change and shift its point.—POINTER. *In layers.*

[Clouds.] In layers. If two strata of clouds appear in hot weather to move in different directions, they indicate thunder.

If, during dry weather, two layers of clouds appear moving in opposite directions, rain will follow.

Clouds floating at different heights show different currents of air, and the upper one generally prevails. If this is north-east, fine weather may be expected; if south-west, rain.

C. L. PRINCE.

Cross wind. If you see clouds going across the wind, there is a storm in the air.

If clouds float at different heights and rates, but generally in opposite directions, expect heavy rains.

Gusts. If there be a cloudy sky, with dark clouds driving fast under higher clouds, expect violent gusts of wind.

Red. Red clouds at sunrise foretell wind; at sunset, a fine day for the morrow.—BACON.

Narrow, horizontal, red clouds after sunset in the west indicate rain before thirty-six hours.

Red clouds in the east, rain the next day.

Black. After black clouds clear weather.

Dark clouds in the west at sunrise indicate rain on that day.

Dull. Clay-coloured and muddy clouds portend rain and wind.

BACON.

Golden. Clouds before sunset of an amber or a gold colour, and with gilt fringes, after the sun has sunk lower, foretell fine weather.

BACON.

Colouring. The wind-gale or prismatic colouring of the clouds is considered by sailors a sign of rain.

Light, delicate, quiet tints or colours, with soft, undefined forms of clouds, indicate and accompany fine weather; but unusual or gaudy hues, with hard, definitely outlined clouds, foretell rain, and probably strong wind.—FITZROY.

Brassy. Brassy-coloured clouds in the west at sunset indicate wind.

Dusky. Dusky or tarnish silver-coloured clouds indicate hail.

HOWARD.

Scud. Small inky-looking clouds foretell rain; light scud clouds driving across heavy masses show wind and rain; but if alone, may indicate wind only.—FITZROY.

Bright and dark.

If clouds be bright,
'Twill clear to-night;
If clouds be dark,
'Twill rain,—do you hark?

He causeth the vapours to ascend from the ends of the earth; He maketh lightnings for the rain; He bringeth the wind out of His treasuries.—PSALM cxxxv. 7.

[Clouds.] Rain.

Clouds above—water below.

Storm.

Generally squalls are preceded, or accompanied, or followed by clouds; but the dangerous white squall of the West Indies is indicated only by a rushing sound and by white wave crests to windward.—FITZROY.

A squall cloud that one can see through or under is not likely to bring or be accompanied by so much wind as a dark, continued cloud extending beyond the horizon.—FITZROY.

Against wind.

If you see a cloud rise against the wind or side wind, when that cloud comes up to you, the wind will blow the same way that the cloud came; and the same rule holds good of a clear place when all the sky is equally thick, except one clear edge.

SHEPHERD OF BANBURY.

Increasing.

A small increasing white cloud about the size of a hand to windward is a sure precursor of a storm.

Behold, there ariseth a little cloud out of the sea, like a man's hand. . . . Prepare thy chariot, and get thee down, that the rain stop thee not. And it came to pass that the heaven was black with clouds and wind, and there was a great rain.

I KINGS xviii. 44, 45.

A small, fast-growing black cloud in violent motion, seen in the tropics, is called the "bull's eye," and precedes the most terrible hurricanes.

Description of.

Sometimes we see a cloud that's dragonish,
A vapour sometimes like a bear or lion,
A towered citadel, a pendent rock,
A forked mountain, a blue promontory
With trees upon't that nod unto the world
And mock our eyes with air.
That which is now a horse, even with a thought,
The rack dislimns and makes it indistinct
As water is in water.

SHAKESPEARE, "ANTONY AND CLEOPATRA."

Bank.

A bench (or bank) of clouds in the west means rain.—SURREY.

Broken.

When small dark clouds (broken nimbi) appear against a patch of blue sky, there will be rain before sunset.

C. L. PRINCE.

Massive.

When you observe greenish-tinted masses of composite cloud collect in the south-east, and remain there for several hours, expect a succession of heavy rains and gales.

C. L. PRINCE.

[CIRRUS.] *Definition.* Parallel, flexuous, or diverging fibres, extensible in any or all directions.—HOWARD.

Common names: Curl Cloud, Mares' Tails, Goat's Hair, etc.
T. FORSTER.

Indicating change. After a long run of clear weather the appearance of light streaks of cirrus cloud at a great elevation is often the first sign of change.—JENYNS.

Showery. Feathery clouds, like palm branches or the *fleur de lis*, denote immediate or coming showers.—BACON.

Indicating wind. Long parallel bands of clouds in the direction of the wind indicate steady high winds to come.

Fine weather. If cirrus clouds dissolve and appear to vanish, it is an indication of fine weather.

Rain. If the cirrus clouds appear to windward, and change to cirro-stratus, it is a sign of rain.

Sheet cirrus. Sheet cirrus occurs with southerly and westerly, but rarely with steady northerly or north-easterly, winds, unless a change to a westerly or southerly quarter is approaching.
HON. F. A. R. RUSSEL.

Rain. In unsettled weather sheet cirrus precedes more wind or rain.

The longer the dry weather has lasted, the less is rain likely to follow the cloudiness of cirrus.

Murky. A large formation of murky white cirrus may merely indicate a backing of wind to an easterly quarter.

Tufted cirrus. This variety is a constant accompaniment of showers in broken weather, and borders the lower clouds with a crown of feathery tufts.

Feathery. If a shower be approaching from the west, it may be seen shooting forth white feathery rays from its upper edge, often very irregular and crooked.

Cirrus of a long, straight, feathery kind, with soft edges and outlines, or with soft, delicate colours at sunrise and sunset, is a sign of fine weather.

Curdled cirrus. This cloud often indicates the approach of bad weather.

Bar or ribbed cirrus. The rapid movement of a cloud, something between cirrus and cirro-cumulus, in distinct dense bars, in a direction at right angles to the length of the bars, is, by itself, a certain sign of a gale of wind. If the bars are sharply defined and close together, the severer will be the storm. Sometimes these bars remind one of the form of a gridiron. The bands move transversely, and generally precede the storm by from twelve to forty-eight hours.—HON. F. A. R. RUSSEL.

Curly wisps and blown-back pieces are not a bad sign. *[Cirrus.] Curly.*

When the tails are turned downwards, fair weather or slight showers often follow. *Tails downwards.*

The harder and more distinct the outline, and the more frequently particular forms are repeated, the worse the result. *Definite.*

Long, hard, greasy-looking streaks, with rounded edges or knobs, whether crossed by fibres at right angles or not, are a sign of storms ; but the storms may be at a distance. *Fibrous.*

Cottony shreds, rounded and clear in outline, indicate dangerous disturbances.

Regular, wavy tufts, with or without cross lines, are bad, especially if the tufts end, not in fibres, but in rounded knobs. *Tufty.*

Feathery cirrus in thick patches at equal distances apart is a sign of storm; so is any appearance of definite waves of alternate sky and cloud; so is any regular repetition of the same form. *Regular.*

Slightly undulating lines of cirrus occur in fine weather; but anything like a deeply indented outline precedes heavy rain or wind. *Undulating.*

Cirrus simply twisted or in zigzag lines of a fibrous character often appears in fine weather; and if not hard, or knotted, or clearly marked off from a serene sky, does not often precede any important change. *Twisted.*

Detached patches of cirrus, like little masses of wool or knotted feathers, in a clear sky, and of unusual figure, moving at more than the average rate, precede disturbances of great magnitude. The rays in straight lines are a good sign. *Detached.*

[The last ten rules are by the HON. F. A. R. RUSSEL.]

Continued wet weather is attended by horizontal sheets of cirrus clouds, which subside quickly, passing into the cirrostratus. *Indicating wet.*

When cirri merge into cirri-strati, and when cumuli increase towards evening and become lower, expect wet weather.

Streaky clouds across the wind foreshow rain.—SCOTLAND.

If cirrus clouds form in fine weather with a falling barometer, it is almost sure to rain.—HOWARD.

These clouds announce the east wind. If their under surface is level, and their streaks pointing upwards, they indicate rain; if downwards, wind and dry weather.—HOWARD. *Rain and wind.*

If the cirrus clouds get lower and denser to leeward, it presages bad weather from the opposite quarter. *Bad weather.*

[Cirrus.] Storms. When the cirrus clouds appear at lower elevations than usual, and with a denser character, expect a storm from the opposite quarter to the clouds.

Pointing upwards. When streamers point upward, the clouds are falling, and rain is at hand; when streamers point downwards, the clouds are ascending, and drought is at hand.

Streaky. When after a clear frost long streaks of cirrus are seen with their ends bending towards each other as they recede from the zenith, and when they point to the north-east, a thaw and a south-west wind may be expected.

Barred. The bar or ribbed cirrus is considered by the Hon. F. A. R. Russel as good a danger signal as that given by a falling barometer.

Weather-head cirrus. In Shetland the name of "weather-head" is given to a band of cirrus passing through the zenith; and they say if it lies north-east to south-west, good weather comes; but if south-east to north-west, a gale is looked for.

[CIRRO-STRATUS.] *Definition.* Horizontal or slightly inclined masses, attenuated towards a part or the whole of their circumference, bent downwards, or undulated, separate, or in groups, or consisting of small clouds having these characters.—HOWARD.

Wind.

If clouds look as if scratched by a hen,
Get ready to reef your topsails then.—NAUTICAL.

Hen's scarts [scratchings] and filly tails
Make lofty ships carry low sails.

Hairy. Comoid cirri, or cirri in detached tufts, called "mares' tails," may be regarded as a sign of wind, which follows, often blowing from the quarter to which the fibrous tails have previously pointed.—T. FORSTER.

Trace in the sky the painter's brush,
Then winds around you soon will rush.

The cloud called "goat's hair" or the "grey mare's tail" forebodes wind.

Ark-like. The form of cloud popularly called "Noah's ark" is also called the "magnetic cirrus," and is said to consist of fine ice crystals, and to be accompanied by magnetic disturbances.

A long stripe of cloud, sometimes called a salmon, sometimes a Noah's ark, when it stretches east and west, is a sign of a storm; but when north and south, of fine weather.

In the Eifel district of the Lower Rhine, on the contrary, they say, when the "cloud ship" turns its head to the south, rain will soon follow.

[Cirro-stratus.] Cloud ship.

When a plain sheet of the wane cloud is spread over a large surface at eventide, or when the sky gradually thickens with this cloud, a fall of steady rain is usually the consequence.

T. FORSTER.

Wane cloud.

In low pressure areas the stripes lie parallel to the isobars (lines of equal barometric pressure), while in high pressure areas the stripes cross the isobars at right angles.

HILDEBRANDSSON.

Direction.

Continuous cirro-strati gathering into unbroken gloom, and also the cloud called "goat's hair," or the "grey mare's tail," presage wind.—SCOTLAND.

Gloomy.

When after a shower the cirro-strati open up at the zenith, leaving broken or ragged edges pointing upwards, and settle down gloomily and compactly on the horizon, wind will follow, and will last for some time.—SCOTLAND.

Indicating wind.

The cirro-stratus precedes winds and rains, and the approach of foul weather may sometimes be inferred from its greater or less abundance, and the permanent character it puts on.

Wind and rain.

If clouds appear high in air in their *white trains*, wind and probably rain will follow.

When ash-coloured masses of cumulo-stratus and cirro-stratus cloud collect over the sea, extending in a line from south-east to south-west, expect rain and probably wind on the second day.—C. L. PRINCE.

Long lines of cirro-strati, extending along the horizon, and *slightly contracted in their centre*, expect heavy rain the following day.—C. L. PRINCE.

Rain.

The cirro-stratus is doubtless the one alluded to by Polonius, in *Hamlet*, as "very like a whale."

Fish-shaped.

The fish (hake) shaped cloud, if pointing east and west, indicates rain; if north and south, more fine weather.

BEDFORDSHIRE.

North and south, the sign of drought;
East and west, the sign of blast.

Light, fleecy clouds in rapid motion, below compact, dark cirro-strati, foretell rain near at hand.—SCOTLAND.

With cirrus.

[Cirro-stratus.] Opening. When after a shower the cirro-strati open up at the zenith, leaving broken or ragged edges pointing upwards, and settle down gloomily and compactly on the horizon, wind will follow, and will last for some time.

Indicating thunder. The waved cirro-stratus indicates heat and thunder.

[CIRRO-CUMULUS.] *Definition.* Small, well-defined, roundish masses increasing from below.

HOWARD.

Commonly called "mackerel sky."

Indicating wind.

Mackerel sky and mares' tails
Make lofty ships carry low sails.

Rain. A mackerel sky denotes fair weather for that day, but rain a day or two after.

Change

Mackerel sky, mackerel sky,
Never long wet and never long dry.

Mackerel clouds in sky,
Expect more wet than dry.

Mackerel scales,
Furl your sails.

A mackerel sky,
Not twenty-four hours dry.

Small. If small white clouds are seen to collect together, their edges appearing rough, expect wind.

Indicating thunder. Before thunder, cirro-cumulus clouds often appear in very dense and compact masses, in close contact.

Curdled.

A curdly sky
Will not leave the earth long dry.

A curdly sky
Will not be twenty-four hours dry.

Direction. When cirro-cumuli appear in winter, expect warm and wet weather. When cirri threads are brushed back from a southerly direction, expect rain and wind.

Packet boys. These clouds are called in Buckinghamshire "packet boys," and are said to be packets of rain soon to be opened.

Small. Small floating clouds over a bank of clouds, sign of rain.

Wandering. In summer we apprehend a future storm when we see little, black, loose clouds lower than the rest, wandering to and fro; when at sunrise we see several clouds gather in the west; and, on the other hand, if these clouds disperse, it speaks fair weather.—OZANAM.

Scattered. Fleecy clouds scattered over the sky denote storms; but clouds which rest upon one another like scales or tiles portend dry and fine weather.—BACON.

A sky dappled with light clouds of the cirro-cumulus form in the early morning generally leads to a fine and warm day.

JENYNS.

[*Cirro-cumulus.*] *Dappled.*

Dappled sky is not for long.—FRANCE.

If woolly fleeces spread the heavenly way,
Be sure no rain disturbs the summer day.

Crowded. Small white clouds, like a flock of sheep, driving north-west, indicate continued fine weather.

Storm. The cirro-cumulus, when accompanied by the cumulo-stratus, is a sure indication of a coming storm.

Outlines. If soft and delicate in outline, it may be followed by a continuance of fine weather; but if dense, abundant, and associated with cirrus, it signifies electrical disturbance and change of wind, often resulting in thunderstorms in summer or gales in winter.

High. High cirro-cumulus commonly appears a few hours or days before thunderstorms. It generally moves with the prevailing surface wind. The harder and more definite the outline, the more unsettled the coming weather. In winter clearly marked, high cirro-cumulus is a sign of bad weather. If the cloud be continuous in long streaks, dense, and with rounded, knobby outlines, stormy weather follows generally within two or three days.

Soft. When cirro-cumulus is seen overhead, if the fleeces gently merge into each other, and the edges are soft and transparent, settled weather prevails; and if the middle part of the fleeces look shadowy, so much the better.

Slow. Cirro-cumulus at a great height and in large masses, moving slowly from north-east, is a sign of the continuance of the wind in that quarter.—HON. F. A. R. RUSSEL.

[CUMULUS.] *Definition.* Convex or conical heaps increasing upwards from a horizontal base—wool-bag clouds.

Stormy. In India, if a cumulus cloud have a stratum of flat cloud above it, a coming storm is indicated.

Piled up. Sometimes the clouds appear to be piled in several tiers or stories, one above the other (Gilbert, *Phys.*, iv. 1, declares that he has sometimes seen and observed five together), whereof the lowest are always the blackest, though it sometimes appears otherwise, as the whiter most attract the sight. Two stories, if thick, portend instant rain (especially if the lower one appear overcharged); many tiers denote a three days' rain.—BACON.

Fleecy.

Refreshing showers or heavier rains are near
When piled in fleecy heaps the clouds appear.

J. LAMB'S "ARATUS."

[Cumulus.] Dark.

If a black cloud eclipse the solar ray,
And sudden night usurp the place of day,
As when th' obtrusive moon's dark orb is seen
Forcing her way the sun and earth between.
J. LAMB'S "ARATUS."

Opening and closing.

If clouds open and close, rain will continue.

Round topped.

A round-topped cloud, with flattened base,
Carries rainfall in its face.

White.

A white loaded cloud, called by the ancients a white tempest, is followed in summer by showers of very small hail, in winter by snow.—BACON.

Wind.

Cumulus clouds high up are said to show that south and south-west winds are near at hand; and stratified clouds low down, that east or north winds will prevail.—SCOTLAND.

Tower-like, indicating rain.

Large irregular masses of cloud, "like rocks and towers," are indicative of showery weather. If the barometer be low, rain is all the more probable.—JENYNS.

When clouds appear like rocks and towers,
The earth's refreshed by frequent showers.

When mountains and cliffs in the clouds appear,
Some sudden and violent showers are near.

When the clouds rise in terraces of white, soon will the country of the corn priests be pierced with the arrows of rain.
ZUÑI INDIANS.

Augmenting.

Before rain these clouds augment in volume with great rapidity, sink to a lower elevation, and become fleecy and irregular in appearance, with their surfaces full of protuberances. They usually also remain stationary, or else sail against the surface wind previous to wet weather.

Banking up.

When the clouds bank up the contrary way to the wind, there will be rain.

If on a fair day in winter a white bank of clouds arise in the south, expect snow.

Water waggons.

The rounded clouds called "water waggons" which fly alone in the lower currents of wind forebode rain.—T. FORSTER.

Diminishing.

When the cumulus clouds are smaller at sunset than they were at noon, expect fair weather.

Wet calm.

The formation of cumulus clouds to leeward during a strong wind indicates the approach of a calm with rain.

Fair weather.

When the cumulus clouds are smaller at sunset than they were at noon, expect fair weather.

If clouds are formed like fleeces, deep and dense, or thick and close towards the middle, the edges being very white, while the surrounding sky is bright and blue, they are of a frosty coldness, and will speedily fall in hail, snow, or rain. [*Cumulus.*] *Indicating hail, snow, or rain.*

And another storm brewing; I hear it sing i' the wind. Yond' same black cloud, yond' huge one, looks like a foul bumbard that would shed his liquor. . . . Yond' same cloud cannot chuse but fall by pailfuls.—SHAKESPEARE'S "TEMPEST." *Storm.*

The pocky * cloud or heavy cumulus, looking like festoons of drapery, forebodes a storm.—SCOTLAND.

In summer or harvest, when the wind has been south for two or three days, and it grows very hot, and you see clouds rise with great white tops like towers, as if one were upon the top of another, and joined together with black on the nether side, there will be thunder and rain suddenly. If two such clouds arise, one on either hand, it is time to make haste to shelter.—SHEPHERD OF BANBURY. *Thunder.*

When cumulus clouds become heaped up to leeward during a strong wind at sunset, thunder may be expected during the night.

Well-defined cumuli, forming a few hours after sunrise, increasing towards the middle of the day, and decreasing towards evening, are indicative of settled weather: if instead of subsiding in the evening and leaving the sky clear they keep increasing, they are indicative of wet.—JENYNS. *Changing.*

The cirro-stratus blended with the cumulus, and either appearing intermixed with the heaps of the latter, or superadding a widespread structure to its base.—HOWARD. [CUMULO-STRATUS.] *Definition.*

When large masses of cumulo-strati cloud collect simultaneously in the north-east and south-west, with the wind east, expect cold rain or snow in the course of a few hours. The wind will ultimately back to north.—C. L. PRINCE. *Collecting.*

When at sea, if the cumulo-stratus clouds appear on the horizon, it is a sign that the weather is going to break up. *On horizon.*

If there be long points, tails, or feathers hanging from the thunder or rain clouds, five or six or more degrees above the horizon, with little wind in summer, thunder may be expected, but the storm will be of short duration. *Tails or feathers.*

A horizontal streak or band of clouds immediately in front of the mountains on the east side of Salt Lake Valley is an *Streak.*

* *Pock,* a bag.

[Cumulo-stratus.] indication of rain within one or two days. When black clouds cover the western horizon, rain will follow soon, and extend to the eastward over the valley.—UNITED STATES.

Striped. If long strips of clouds drive at a slow rate high in air, and gradually become larger, the sky having been previously clear, expect rain.

[NIMBUS.] *Definition.* A rain cloud—a cloud or system of clouds from which rain is falling. It is a horizontal sheet over which the cirrus spreads, while the cumulus enters it laterally and from beneath.

Rain. By watering He wearieth the thick cloud.—JOB xxxvii. 11.

Prophet clouds. When scattered patches or streaks of nimbus come driving up from the south-west, they are called by the sailors "prophet clouds," and indicate wind.

Storm. If a little cloud suddenly appear in a clear sky, especially if it come from the west, or somewhere in the south, there is a storm brewing.—BACON.

[STRATUS.] *Definition.* A widely extended, continuous, horizontal sheet, increasing from below.—HOWARD.

Fine. These clouds have always been regarded as the harbingers of fine weather, and there are few finer days in the year than when the morning breaks out through a disappearing stratus cloud.

Night. A stratus at night, with a generally diffused fog the next morning, is usually followed by a fine day, if the barometer be high and steady. If the barometer keep rising, the fog may last all day; if the barometer be low, the fog will probably turn to rain.—JENYNS.

On mountains. When mountains extend north and south, if fog or mist comes from the west, expect fair weather. If mist comes from the top of mountains, expect rain in summer, snow in winter.
APACHE INDIANS.

Fair weather. Thin, white, fleecy, broken mist, slowly ascending the sides of a mountain whose top is uncovered, predicts a fair day.
SCOTLAND.

If towers to sight, uncapt, the mountain's head,
While on its base a vapoury veil is spread,
[Fair weather follows].—J. LAMB'S "ARATUS."

On hills. If mist rise to the hilltops and there stay, expect rain shortly.

When the mist comes from the hill,
Then good weather it doth spill;
When the mist comes from the sea,
Then good weather it will be.

When the mist creeps up the hill,
Fisher, out and try your skill;
When the mist begins to nod,
Fisher, then put past your rod.—KIRKCUDBRIGHT.

[Stratus.] On hills.

Misty clouds, forming or hanging on heights, show wind and rain coming, if they remain, increase, or descend. If they rise or disperse, the weather will improve.—FITZROY.

Rising and falling.

Clouds upon hills, if rising, do not bring rain; if falling, rain follows.

When the clouds on the hilltops are thick and in motion, rain to the south-west is regarded as certain to follow.
SCOTLAND.

Thick.

When it gangs up i' fops,*
It'll fa' down i' drops.—NORTH COUNTRY.

Small.

When mountains and hills appear capped by clouds that hang about and embrace them, storms are imminent.—BACON.

Hanging.

When the clouds go up the hill,
They'll send down water to turn a mill.
HAMPSHIRE.

Ascending.

When the clouds are upon the hills,
They'll come down by the mills.

When the Pendle's Head is free from clouds, the people thereabout expect a halcyon day, and those on the banks of the Can (or Kent) in Westmoreland can tell what weather to look for from the voice of its falls.

For when they to the north the noise do easliest hear,
They constantly aver the weather will be clear.
And when they to the south, again they boldly say
It will be clouds or rain the next approaching day.
DRAYTON'S "POLYOLBION."

Pendle's Head.

When Wolsonbury has a cap,
Hurstpierpoint will have a drap.—SUSSEX.

Wolsonbury.

Clouds on Ross-shire Hills mean rain at Ardersier on the south-east of the Moray Frith.

Ross.

Clouds on Bell Rock Light mean rain at Arbroath.

Bell Rock.

Clouds on Orkney Isles mean rain at Cape Wrath.

Orkney.

Clouds on Kilpatrick Hills mean rain at Eaglesham, in Renfrewshire.

Kilpatrick Hills.

Clouds on Ailsa Craig mean rain at Cumbrae.

Ailsa Craig.

* Small clouds on hills.

[Stratus.] *Cape Town.* Sailors say it is a sign of bad weather when the "tablecloth" (a cloud so called) is spread on Table Mountain.

Bever.

If Bever hath a cap,
You churls of the vale look to that.
LEICESTERSHIRE.

Skiddaw.

If Skiddaw hath a hat,
Scruffel wots full well of that.—CUMBERLAND.

When Skiddaw hath a cap,
Criffel wots fu' well of that.

Heavy clouds on Skiddaw, especially with a south wind, the farmer of Kirkpatrick Fleming looks on as an indication of coming rain.

[*Note.*—Skiddaw lies to the south of the place.]

Moncayo.

When Moncayo and Guara have their white caps on,
It is good for Castile and better for Aragon.—SPANISH.

Traprain.

When Traprain puts on his hat,
The Lothian lads may look to that.
HADDINGTONSHIRE.

Ruberslaw.

When Ruberslaw puts on his cowl,
The Dunion on his hood,
Then a' the wives of Teviotside
Ken there will be a flood.—ROXBURGHSHIRE.

[Also said of Craigowl and Collie Law in Forfarshire, substituting "Lundy lads" for "the wives of Teviotside."—ROBERT CHAMBERS.]

Falkland Hill, Lomond Range.

When Falkland Hill puts on his cap,
The Howe o' Fife will get a drap;
And when the Bishop draws his cowl,
Look out for wind and weather foul.

Cheviot.

When Cheviot ye see put on his cap,
Of rain ye'll have a wee bit drap.—SCOTLAND.

Largo Law.

When Largo Law puts on his hat,
Let Kellie Law beware of that;
When Kellie Law gets on his cap,
Largo Law may laugh at that.—SCOTLAND.

[*Note.*—Largo Law is to the south-west of Kellie Law.]

Cairnsmore.

When Cairnsmore wears a hat,
The Macher's Rills may laugh at that.

[*Note.*—Cairnsmore is north-north-east of Macher's Rills, Wigtownshire, Scotland.]

Corsancone. If Corsancone put on his cap, and the Knipe be clear, it will rain within twenty-four hours.

[*Note.*—This is a sign which it is said never fails. Corsancone Hill is to the east and the Knipe to the south-west of the New Cumnock districts, where the proverb is current.]

A cloud on Sidlaw Hills foretells rain to Carmylie. *[Stratus.]*
" Bin Hill " " " Cullen. *Scotch Hills.*
" Paps of Jura " " " } Gigha and
" Mull of Kintyre " " " } Cara.

The rolling of clouds landward and their gathering about the summit of Criffel is regarded as a sign of foul weather in Dumfries and Kirkpatrick Fleming, and intervening parishes. *Criffel.*
[*Note.*—Criffel is to the south-west of the place.]

There is a high wooded hill above Lochnaw Castle; *Craighill.*
Take care when Lady Craighill puts on her mantle.
The Lady looks high and knows what is coming;
Delay not one moment to get under covering.

[*Note.*—The hill lies to the north-west of the district where this doggerel is quoted.]

If Riving Pike do wear a hood, *Riving Pike.*
Be sure the day will ne'er be good.
LANCASHIRE.

A cloud, called the "helm cloud," or "helm bar," hovering about the hilltops for a day or two, is said to presage wind and rain.—YORKSHIRE. *Helm cloud.*

If Roseberry Topping wears a cap, *Roseberry Topping.*
Let Cleveland then beware of a rap.

When Bredon Hill puts on his hat, *Bredon Hill.*
Ye men of the vale, beware of that.
WORCESTERSHIRE.

When Hall Down has a hat, *Hall Down.*
Let Kenton beware of a skat [shower].

When Lookout Mountain has its cap on, it will rain in six hours.—UNITED STATES. *Lookout.*

Mists.

If mists and fogs ascend and return upwards, they denote rain; and if this take place suddenly, so that they appear to be sucked up, they foretell winds; but if they fall and rest in the valleys, it will be fine weather.—BACON. MISTS. *Disappearing.*

Wherever there is a plentiful generation of vapours, and that at certain times, you may be sure that at those times periodical winds will arise.—BACON. *Vapours and winds.*

White mist in winter indicates frost.—SCOTLAND. *White.*

[Mist.] Black. Black mist indicates coming rain.

Mist and rain. Mists above, water below.—SPANISH.

In low ground. If mists rise in low ground and soon vanish, expect fair weather.—SHEPHERD OF BANBURY.

River. A white mist in the evening, over a meadow with a river, will be drawn up by the sun next morning, and the day will be bright. Five or six fogs successively drawn up portend rain.

Rising. Where there are high hills, and the mist which hangs over the lower lands draws towards the hills in the morning, and rolls up to the top, it will be fair; but if the mist hangs upon the hills, and drags along the woods, there will be rain.—REV. W. JONES.

In the evenings of autumn and spring, vapour arising from a river is regarded as a sure indication of coming frost.
SCOTLAND.

Spreading.
Mists dispersing on the plain,
Scatter away the clouds and rain;
But when they rise to the mountain tops,
They'll soon descend in copious drops.

Misty morning. Three foggy or misty mornings indicate rain.—OREGON.

HAZE. Haze and western sky purple indicate fair weather.

Hazy weather is thought to prognosticate frost in winter, snow in spring, fair weather in summer, and rain in autumn.
SCOTLAND.

A sudden haze coming over the atmosphere is due to the mixing of two currents of unequal temperatures: it may end in rain, or in an increase of temperature; or it may be the precursor of a change, though not immediate.—JENYNS.

Clearing. When the landscape looks clear, having your back towards the sun, expect fine weather; but when it looks clear with your face towards the sun, expect showery, unsettled weather.
C. L. PRINCE.

Dew.

DEW.
Evening.
The dews of the evening industriously shun;
They're the tears of the sky for the loss of the sun.

If the dew lies plentifully on the grass after a fair day, it is a sign of another. If not, and there is no wind, rain must follow.—REV. W. JONES.

When in the morning the dew is heavy and remains long on the grass, when the fog in the valleys is slowly dispersed and lingers on the hillsides, when the clouds seem to be taking a higher place, and when a few loose cirro-strati float gently along, serene weather may be expected for the greater part of that day.—SCOTLAND. *Dew and fog.*

If in clear summer nights there is no dew, expect rain next day.—C. L. PRINCE. *Night.*

Dew is an indication of fine weather; so is fog.—FITZROY. *Fine weather.*

Dew is produced in serene weather and in calm places. *Calm.*
ARISTOTLE.

If the dew is evaporated immediately upon the sun rising, rain and storm follow in the afternoon; but if it stays and glitters for a long time after sunrise, the day continues fair. *Dispersing.*
DE QUINCEY'S "NOTE TO ANALECTS FROM RICHTER."

If there is a profuse dew in summer, it is about seven to one that the weather will be fine.—E. J. LOWE. *Profuse.*

With dew before midnight,
The next day will sure be bright. *Evening.*

During summer a heavy dew is sometimes followed by a southerly wind in the afternoon. *South wind.*

If there is a heavy dew, it indicates fair weather; no dew, it indicates rain. *Heavy.*

If nights three dewless there be,
'Twill rain you're sure to see. *Rain.*

When the dew is seen shining on the leaves, the mist rolled down from the mountain last night.—ZUÑI INDIANS. *Mountain.*

When there is no dew at such times as usually there is, it foreshoweth rain.—WING, 1649. *No dew.*

Fog.

When the fog falls, fair weather follows; when it rises, rain ensues. FOG. *Falling.*

In the Mississippi valley, when fogs occur in August, expect fever and ague in the following fall. *August.*

If there be a damp fog or mist, accompanied by wind, expect rain. *Damp.*

Light fog passing under sun from south to north in the morning indicates rain in twenty-four or forty-eight hours. *Light.*

If there be continued fog, expect frost.—UNITED STATES. *With frost.*

[Fog.] Hunting. When the fog goes up the mountain, you may go hunting; when it comes down the mountain, you may go fishing. In the former case it will be fair, in the latter it will rain.

Change. Fogs are signs of a change.

Winter. Heavy fog in winter, when it hangs below trees, is followed by rain.

Sea. Fog from seaward, fair weather; fog from landward, rain.

NEW ENGLAND.

Sea and hills.

A fog from the sea
Brings honey to the bee;
A fog from the hills
Brings corn to the mills.

PEMBROKESHIRE.

Hanging. When with hanging fog smoke rises vertically, rain follows.

Sky.

SKY.

Clear. A very clear sky without clouds is not to be trusted, unless the barometer be high.—JENYNS.

Foul. So foul a sky clears not without a storm.

SHAKESPEARE'S "KING JOHN."

Hazy. One of the surest signs of rain with which I am acquainted is that of the sky assuming an almost colourless appearance in the direction of the wind, especially if lines of dark or muddy cirro-strati lie above and about the horizon and the milkiness gradually becomes muddy.—E. J. LOWE.

Greenish. If the sky is of a deep, clear blue or a sea-green colour near the horizon, rain will follow in showers.

In winter, when the sky at midday has a greenish appearance to the east or north-east, snow and frost are expected.

SCOTLAND.

When the sky in rainy weather is tinged with sea-green, the rain will increase; if with deep blue, it will be showery.

REV. W. JONES.

Blue space. A small cloudless place in the north-east horizon is regarded both by seamen and landsmen as a certain precursor of fine weather or a clearing up.—SCOTLAND.

Enough blue sky in the north-west to make a Scotchman a jacket is a sign of approaching clear weather; and the same is said satirically of a Highlandman's "breeks."

When as much blue is seen in the sky as will make a Dutchman's jacket (or a sailor's breeches), the weather will clear.

[Sky.] *Clear.* Clear in the south beguiled the cadger.—SCOTLAND.

Grey. If there be a dark grey sky with a south wind, expect frost.

Dark. If the sky become darker, without much rain, and divides into two layers of clouds, expect sudden gusts of wind.

Colours. A dark, gloomy blue sky is windy; but a light, bright blue sky indicates fine weather. When the sky is of a sickly looking, greenish hue, wind or rain may be expected.
FITZROY.

Carlisle. From Dumfries to Gretna a lurid, yellowish sky in the east or south-east is called a Carlisle or Carle sky, and is regarded as a sure sign of rain.—SCOTLAND.

The Carle sky
Keeps not the head dry.

Reflecting. In Kincardine of Monteith, and in all that district, the reflection from the clouds, of the furnaces of the Devon and Carron works (to the east) foretells rain next day.—SCOTLAND.

The glare of the distant Ayrshire ironworks being seen at night from Cumbrae on Rothesay, rain is expected next day.
SCOTLAND.

Air.

AIR.

Undulation. Much undulation in the air on a hot day in May or June foretells cold.—SCOTLAND.

Clearness. The farther the sight the nearer the rain.

When the distant hills are more than usually distinct, rain approaches.

The cliffs and promontories of the shore appear higher and the dimensions of all objects seem larger when the south-east wind is blowing.—ARISTOTLE.

Lizard Point.
When the Lizard is clear,
Rain is near.—CORNWALL.

Shipping. The unusual elevation of distant coasts, masts of ships, etc., particularly when the refracted images are inverted, are known to be frequent foreboders of stormy weather.

Isle of Wight. When the Isle of Wight is seen from Brighton or Worthing, expect rain soon.

Mirage. A mirage is followed by a rain.—NEW ENGLAND.

Sound.

SOUND.

A good hearing day is a sign of wet.

There is a sound of abundance of rain.—ELIJAH.

Bells. The ringing of bells is heard at a greater distance before rain; but before wind it is heard more unequally, the sound coming and going, as we hear it when the wind is blowing perceptibly.—BACON.

[Sound.] In air. A sound from the mountains, an increasing murmur in the woods, and likewise a kind of crashing noise in the plains, portend winds. An extraordinary noise in the sky when there is no thunder is principally due to winds.—BACON.

Air.

A sound in air presaged approaching rain,
And beast to covert scud across the plain.
THOMAS PARNELL.

On shore. The shores sounding in a calm, and the sea beating with a murmur or an echo louder and clearer than usual, are signs of wind.—BACON.

The calling of the sea. A murmuring or roaring noise, sometimes heard several miles inland during a calm, in the direction from which the wind is about to spring up.

Pons-an-dane.

When Pons-an-dane calls to Larrigan river,
There will be fine weather;
But when Larrigan calls to Pons-an-dane,
There will be rain.—CORNWALL.

[*Note.*—Streams entering the sea north-east and south-west of Penzance, about one mile and a half apart, Pons-an-dane being north-east.—RICHARD EDMONDS.]

Rosehearty. If the "sang" of the sea is heard coming from the west by the fishermen of Rosehearty in the morning, when they get out of bed to examine the state of the weather, whether favourable or unfavourable to fishing, it is regarded as an indication of fine weather for the day, and accordingly they sometimes go farther to sea.—WALTER GREGOR IN "FOLK-LORE JOURNAL."

Fortingal. In Fortingal (Perthshire), if in calm weather the sound of the rapids on the Lyon is distinctly heard, and if the sound descends with the stream, rainy weather is at hand; but if the sound goes up the stream, and dies away in the distance, it is a sign of continued dry weather, or a clearing up, if previously thick.

Travelling.

Sound travelling far and wide,
A stormy day will betide.

Monzie. When the people of Monzie (Perthshire) hear the sound of the waterfalls of Shaggie or the roar of the distant Turret clearly and loudly, a storm is expected; but if the sound seems to recede from the ear till it is lost in the distance, and if the weather is thick, a change to fair may be looked for speedily.

Dysart. In the collieries about Dysart, and in some others, it is thought by the miners that before a storm of wind a sound not unlike that of a bagpipe or the buzz of the bee comes from the mineral, and that previous to a fall of rain the sound is more subdued.—SIR A. MITCHELL. [*Sound.*]

Sounds are heard with unusual clearness before a storm. The railway whistle, for instance, seems remarkably shrill. *Whistle.*

Tide, etc.

Showers occur more frequently at the turn of the tide. *Tide.*

Storms burst as the tide turns.—SOUTH ATLANTIC COAST. *Storms.*

If, after the first ebb of the tide, it flows again for a little while, a storm approaches.—SCOTCH COAST.

The sea swelling silently and rising higher than usual in the harbour, or the tide coming in quicker than ordinary, prognosticates wind.—BACON. *Wind.*

If it raineth at tide's flow,
You may safely go and mow;
But if it raineth at the ebb,
Then, if you like, go off to bed. *Ebb and flow.*

Rain is likely to commence on the turn of the tide. *Turn.*

In threatening weather it is more apt to rain at the turn of the tide, especially at high water. *Rain.*

If, during the absence of wind, the surface of the sea becomes agitated by a long rolling swell, a gale may be expected. This is well known to seamen. *Swell.*

On the west coast a heavy surf is considered the sure forerunner of a storm; while on the east a peculiar ripple, called a "twine," along the surface is known to precede a gale from the south-east. *Surf.*

Just before a storm the sea heaves and sighs.—FITZROY. *Sigh of sea.*

A river flood,
Fishers' good.—SPANISH. *Flood.*

If the river Tweed rise without rain, it foretells the same within twelve hours. *Tweed.*

When the surface of the sea in harbour appears calm, and yet there is a murmuring noise within it, although there is no swell, a wind is coming.—BACON. *Sea appears calm.*

When the foam of the sea retreats or goes out ("works oot"), it is said to be "leukin for mair"; and more stormy weather is looked upon as at hand at Rosehearty. *Sea foam.*

WALTER GREGOR IN "FOLK-LORE JOURNAL."

Sea foam. Glittering foam (called "sea lungs") in a heavy sea foretells that the storm will last many days.—BACON.

Bubbles. If foam, white circles of froth, or bubbles of water, appear here and there on a calm and smooth sea, they prognosticate wind. If these signs be more striking, they denote severe storms.—BACON.

River foam. Much foam in a river foretells a storm.—SCOTLAND.

Phosphorescence of waves. When the phosphorescence of the sea is seen during a dark night on the breast of the roll, or on the water as it breaks on the rocks, it is looked upon as an indication of coming foul weather.—"FOLK-LORE JOURNAL."

Waterspouts. Waterspouts are not produced in cold weather.—ARISTOTLE.

Sudden changes of temperature. A sudden increase in the temperature of the air sometimes denotes rain; and again a sudden change to cold sometimes forebodes the same thing.—BACON.

A sudden and extreme change of temperature of the atmosphere, either from heat to cold, or cold to heat, is generally followed by rain within twenty-four hours.—DALTON.

Temperature. A high temperature, with a high dew-point, and the wind south or south-west, is likely to produce a thunderstorm. If the mercury falls much previous to the storm, the latter is likely to be succeeded by a change of weather. Sometimes heavy thunderstorms take place overhead without any fall of the mercury: in this case a reduction of temperature does not usually follow.—BELVILLE.

"Weather breeders." Fine warm days are called "weather breeders."

Damp heat. What is called "foul air," accompanied by the cheeping of small birds, foreshows a gale from the south or south-east.
KINTYRE.

Rain.

RAIN. Rain comes from a mass of vapour which is cooled.
ARISTOTLE.

Mountains. Mountains cool the uplifted vapour, converting it again into water.—ARISTOTLE.

Wind. When God wills, it rains with any wind.—SPANISH.

Calm.
More rain, more rest;
Fine weather not the best.—NAUTICAL.

Some rain, some rest;
Fine weather isn't always best.

Changes.
No one so surely pays his debt
As wet to dry and dry to wet.—WILTSHIRE.

[Rain.] *North-east.*

With the rain of the north-east comes the ice fruit [hail].
ZUÑI INDIANS.

Rain from the north-east in Germany continues three days.

East.

Rain from the east,
Two days at least.

South.

Rain from the south prevents the drought;
But rain from the west is always best.

Rain which sets in with a south wind on the North Pacific coast will probably last.

If it begin to rain from the south, with a high wind, for two or three hours, and the wind falls, but the rain continues, it is likely to rain twelve hours or more, and does usually rain till a north wind clears the air. These long rains seldom hold above twelve hours, or happen above once a year.
SHEPHERD OF BANBURY.

Rain with south or south-west thunder brings squalls on successive days.

West.

When rain comes from the west, it will not last long.
UNITED STATES.

Squalls.

When rain squalls break to the westward, it is a sign of foul weather. When they break to the leeward, it is a sign of fair weather.—NORTH-EAST COAST, UNITED STATES.

Short.

The faster the rain, the quicker the hold up.—NORFOLK.

Long foretold.

Rain long foretold, long last;
Short notice, soon past.

Small showers.

Small showers last long, but sudden storms are short.
SHAKESPEARE'S "RICHARD II."

Morning.

Rain before seven,
Lift before eleven.

If rain begins at early morning light,
'Twill end ere day at noon is bright.

Morning rains are soon past.—FRANCE.

Rules.

The following rules are believed in by some with respect to the times of rain:—
If rain commences before daylight, it will hold up before 8 a.m.; if it begins about noon, it will continue through the afternoon; if it commences after 9 p.m., it will rain the next day; if it clears off in the night, it will rain the next day; if the wind is from the north-west or south-west, the storm will be short; if from the north-east, it will be a hard one; if from the north-west a cold one, and from the south-west a warm one. If it ceases after 12 a.m., it will rain next day; if it ceases before 12 a.m., it will be clear next day. If it begins about 5 p.m., it will rain through the night.

[Rain.] Custom. In Burmah the inhabitants have a custom of pulling a rope to produce rain. A rain party and a drought party tug against each other, the rain party being allowed the victory, which in the popular notion is generally followed by rain.

"FOLK-LORE JOURNAL," VOL. I., P. 214.

Before sunrise. If it begin to rain an hour or two before sunrising, it is likely to be fair before noon, and so continue that day; but if the rain begin an hour after sunrising, it is likely to rain all that day, except the rainbow be seen before it rains.

SHEPHERD OF BANBURY.

Rain a short time before sunrise will be followed at least by a fine afternoon; but rain soon after sunrise, generally by a wet day.

Dew. If the rain falls on the dew, it will fall all day.—BERGAMO.

Drizzle. A fall of small drizzling rain, especially in the morning, is a sure sign of wind to follow.—NEWHAVEN.

Midnight. If it rain at midnight with a south wind, it will generally last above twelve hours.

Rain and wind. After rains, the wind most often blows in the places where the rain falls, and winds often cease when rain begins to fall.

ARISTOTLE.

A hasty shower of rain falling when the wind has raged some hours, soon allays it.—POINTER.

Small rain abates high wind.—FRANCE.

Marry the rain to the wind, and you have a calm.

Three days' rain.

The wise have in mind the three days' wind,
That foretells the stormy rain;
And to them the care how they then shall fare
Is about the thought of gain.—CARY'S "PINDAR."

Small. A small rain may allay a great storm.—T. FULLER.

Sudden. Sudden rains never last long; but when the air grows thick by degrees, and the sun, moon, and stars shine dimmer and dimmer, then it is likely to rain six hours usually.

SHEPHERD OF BANBURY.

From mountains. They are wet with the showers of the mountains.—JOB xxiv. 8.

Uncertain.

It rains by planets.

To talk of the weather, it's nothing but folly;
For when it's rain on the hill, it may be sun in the valley.

Sunshine. If it rains when the sun shines, it will rain the next day.

If it rains while the sun is shining, the devil is beating his grandmother. He is laughing, and she is crying.

After rain comes sunshine. *[Rain.]*

Sunshine and shower, rain again to-morrow. *Sunshine.*

If it rain when the the sun shines, it will surely rain the next day about the same hour.—SUFFOLK.

A sunshiny shower
Never lasts half an hour.—BEDFORDSHIRE.

Sunshiny rain
Will soon go again.—DEVONSHIRE.

When *fine*, take your unbrella. *Umbrella.*
When *raining*, please yourself.—DR. JOHNSON.

If short showers come during dry weather, they are said to "harden the drought" and indicate no change.—SCOTLAND. *Showers short.*

There is usually fair weather before a settled course of rain. FITZROY. *Preceded by fair weather.*

A foot deep of rain *Rain and snow.*
Will kill hay and grain;
But three feet of snow
Will make them come mo [more].—DEVONSHIRE.

If hail appear after a long course of rain, it is a sign of clearing up.—SCOTLAND. *Followed by hail.*

"Agree between yourselves," quoth Arlotto, "and I will make it rain."—ITALIAN. *Desired.*

Who soweth in rain, he shall reap it with tears.—TUSSER. *Sowing.*

When it rains, they say in Amorgos, "God is emptying His bowl," the prevalent idea being that God, like Zeus of antiquity, has a bowl or receptacle full of water, which He shakes, and then clouds come out; these fall on the earth as rain or snow.—T. BENT (GREECE). *Bowl of Zeus.*

Though it rains, do not neglect to water.—SPANISH. *Watering.*

After great droughts come great rains.—DUTCH. *Drought..*

Wet continues, if the ground dries up too soon. *Continued.*

Rainbow.

The old Norsemen called the rainbow "The bridge of the gods."—C. SWAINSON. RAINBOW.

A rainbow can only occur when the clouds containing or depositing the rain are opposite to the sun; and in the evening the rainbow is in the east, and in the morning in the west; and as our heavy rains in this climate are usually brought by the westerly wind, a rainbow in the west indicates that the bad weather is on the road, whereas the rainbow in the east proves that the rain in these clouds is passing from us.—SIR HUMPHRY DAVY IN "SALMONIA." *East and west.*

[Rainbow.] In cloud. When a rainbow is formed in an approaching cloud, expect a shower; but when in a receding cloud, fine weather.

C. L. PRINCE.

In spring. A rainbow in spring indicates fair weather for twenty-four hours.

In wind's eye. When a rainbow appears in wind's eye, rain is sure to follow.

Windward.

Rainbow to windward, foul fall the day;
Rainbow to leeward, damp runs away.—NAUTICAL.

Fair and foul. If a rainbow appear in fair weather, foul will follow; but if a rainbow appear in foul weather, fair will follow.

Morning and evening. Rainbow in morning shows that shower is west of us, and that we shall probably get it. Rainbow in the evening shows that shower is east of us, and is passing off.—UNITED STATES.

The weather's taking up now,
For yonder's the weather gaw; *
How bonny is the east now!
Now the colours fade awa'.—GALLOWAY.

A dog in the morning,
Sailor, take warning;
A dog in the night
Is the sailor's delight.

[A sun dog, in nautical language, is a small rainbow near the horizon.—ROPER.]

A rainbow in the morn, put your hook in the corn;
A rainbow in the eve, put your hook in the sheave.

CORNWALL.

If there be a rainbow in the eve,
It will rain and leave;
But if there be a rainbow in the morrow,
It will neither lend nor borrow.

A rainbow in the morning
Is the shepherd's warning;
A rainbow at night
Is the shepherd's delight.

The rainbow in the marnin'
Gives the shepherd warnin'
To car' his gurt cwoat on his back;
The rainbow at night is the shepherd's delight,
For then no gurt cwoat will he lack.

WILTSHIRE.

* Fragmentary rainbow.

[Rainbow.] Direction.

A rainbow in the west brings dew and light showers; a rainbow in the east promises fair weather.—SENECA.

[The same author mentions a rainbow in the south, though he does not say how this can be.]

High.

When the rainbow does not reach down to the water, clear weather will follow.

Low.

A bow low down on the mountains is a bad sign for the crops. If seen at a great distance, it indicates fair weather.

Colours.

When a perfect rainbow shows only two principal colours, which are generally red and yellow, expect fair weather for several days.—C. L. PRINCE.

Blue.

If a blue colour should predominate, the air is clearing.

Various.

These colours [of the rainbow] are almost the only ones which the painters cannot reproduce. They try to obtain some by various mixtures; but the red, the green, and the violet cannot be the result of a mixture. And it is these colours which we see in a rainbow.—ARISTOTLE.

[Red, green, and violet are now again considered as the true primary colours.—R. I.]

Indications.

If the green be large and bright in the rainbow, it is a sign of continued rain. If red be the strongest colour, there will be rain and wind together. After much wet weather the rainbow indicates a clearing up. If the bow disappears all at once, there will follow serene and settled weather. The bow in the morning, rain will follow; if at noon, heavy rain; if at night, fair weather. The appearance of double or triple bows indicates fair weather for the present, but heavy rains soon.

Double.

Aristotle knew of the two rainbows having the colours in the reverse order, as he speaks of the red being outside the inner bow and inside the outer one. He also says there are never more than two bows.

Broad.

When the rainbow is broad, with the prismatic colours very distinct, and green or blue predominating, expect much rain the succeeding night. If the red colour is conspicuous and the last to disappear, expect both rain and wind.—C. L. PRINCE.

Prevailing colours.

The peasants of Anaphi are said to know how to foretell the crops by the colours of the rainbows. If red prevails, the crop of grapes will be abundant; if green, that of olives; if yellow, that of corn. A rainbow in the morning denotes luck; in the evening, woe. It is called the "nun's girdle."

T. BENT (GREECE).

After drought.

The rainbow, after a long drought, is the precursor of a decided change to wet weather; and it happens also that a perfect bow, after an unsettled time, is a precursor of fair weather.—C. L. PRINCE.

[Rainbow.] Suddenly appearing. If the rainbow forms and disappears suddenly, the prismatic colours being but slightly discernible, expect fair weather next day.—C. L. PRINCE.

Frost.

FROST.

If hoar-frost come on mornings twain,
The third day surely will have rain.

Not permanent. Hoar-frost and gipsies never stay nine days in a place.

A white frost never lasts more than three days.

Vines. Hoar-frost is good for vines, but bad for corn.—FRANCE.

Bearded frost. Bearded frost, forerunner of snow.

Formation. When it is a cloud which is frozen, snow results; when it is a vapour only, then it produces hoar-frost only.—ARISTOTLE.

Rain. Rain is sure to follow after frost that melts before the sun rises.

Frost and thaw. A very heavy white frost in winter is followed by a thaw.
UNITED STATES.

Rain. Hoar-frost indicates rain.

When the frost gets into the air, it will rain.

Three frosts in succession are a sign of rain.

Light. Light or white frosts are always followed by wet weather, either the same day or three days after.

Late. If the first frost occurs late, the following winter will be mild, but weather variable. If the first frost occurs early, it indicates a severe winter.

Early. Early frosts are generally followed by a long and hard winter.

Storm. Three white frosts and then a storm.

Duration. Six months from last frost to next frost.—SOUTHERN STATES.

Black. Black frost indicates dry, cold weather.

A black frost is a long frost.

Foul. Frosts end in foul weather.

Heavy. Heavy frosts are generally followed by fine, clear weather.
UNITED STATES.

Heavy frosts bring heavy rains; no frost, no rain.
CALIFORNIA.

Frost and fog. He that would have a bad day maun gang out in a fog after a frost.—SCOTCH.

Frost and mists. In the change from frost to open weather, or from open weather to frost, commonly great mists.—BACON.

During frosty weather, the dissolution of mist, and the appearance of small detached cirro-cumulus clouds in the upper air, indicate a thaw. *[Frost.] Mist.*

Signs of frost breaking up :— *Breaking.*

The sun looking waterish at rising.

The sun setting in bluish clouds, and casting reflected rays into them.

The stars looking dull, and the moon's horns blunted, aid the frost to depart.

Quick thaw, long frost.—OLD ANGLO-SAXON. *Long.*

A thaw after a frost doth greatly rot and mellow the ground. BACON. *Beneficial.*

In frosty weather the stars appear clearest and most sparkling.—BACON. *Stars.*

Hail.

Hail brings frost in the tail. HAIL.

Hail is rare in winter.—ARISTOTLE. *Winter.*

Hail is formed in the clouds, and never in the lower mists. ARISTOTLE. *Formation.*

A hailstorm by day denotes a frost at night. *Hailstorm.*

Snow.

Snow cherisheth the ground and anything sowed in it. BACON. SNOW. *Beneficial.*

Corn is as comfortable under the snow as an old man is under his fur cloak.—RUSSIA.

Much snow, much hay.—SWEDEN. *Hay.*

The snows dissolve fastest upon the sea coasts, yet the winds are counted the bitterest from the sea, and such as trees will bend from.—BACON. *Dissolving.*

In winter, during a frost, if it begin to snow, the temperature of the air generally rises to 32° (or near it), and continues there whilst the snow falls; after which, if the weather clear up, expect severe cold.—DALTON. *Temperature.*

Nae hurry wi' your corns,
Nae hurry wi' your harrows;
Snaw lies ahint the dyke;
Mair may come and fill the furrows.—SCOTLAND. *Harrowing.*

[Snow.] *Cloudy.* It takes three cloudy days to bring a heavy snow.
NEW ENGLAND.

Healthy. The more snow, the more healthy the season.
JOHN AYERS (SANTA FÉ).

Snow is generally preceded by a general animation of man and beast, which continues until after the snowfall ends.
UNITED STATES.

Flakes. If the snowflakes increase in size, a thaw will follow.

Wet. If the first snow sticks to the trees, it foretells a bountiful harvest.

November. If the snow remains on the trees in November, they will bring out but few buds in the spring.—GERMAN.

Crops. A heavy fall of snow indicates a good year for crops, and a light fall the reverse.—DR. JOHN MENUAL.

New moon. Snow coming two or three days after new moon will remain on the ground some time, but that falling just after new moon will soon go off.

Moon. As many days old as the moon is at the first snow, there will be as many snows before crop-planting time.

Mud. When snow falls in the mud, it remains all winter.

Last snow. The number of days the last snow remains on the ground indicates the number of snowstorms which will occur during the following winter.

Dry. If the snow that falls during the winter is dry, and is blown about by the wind, a dry summer will follow. Very damp snow indicates rain in the spring.

Lying.

When the snow falls dry, it means to lie;
But flakes light and soft bring rain oft.

When now in the ditch the snow doth lie,
'Tis waiting for more by-and-bye.

Ice.

ICE. If the ice crack much, expect frost to continue.

Thunder and Lightning.

THUNDER AND LIGHTNING.

First thunder.

The thunderstorms of the season will come from the same quarter as the first one.

First thunder in winter or spring indicates rain and very cold weather.—DR. JOHN MENUAL.

After the first thunder comes the rain.—ZUÑI INDIANS.

According to the direction from which comes the first thunder in spring, the Zuñi Indians reckon the coming season. If the thunder be in the north, they say that the bear in his cave has stretched out his left leg; if in the east, that he has stretched out his right arm, and that the winter is over; if in the south, that he has merely stretched out his right leg; or if in the west, his left arm.—MAJOR DUNWOODY.

The first thunder of the year awakes
All the frogs and all the snakes.

Spring thunder.

If there be showery weather, with sunshine and increase of heat, in the spring, a thunderstorm may be expected every day, or at least every other day.

Summer thunder.

Thunder and lightning in the summer show
The point from which the freshening breeze will blow.
J. LAMB'S "ARATUS."

Heat and thunder.

Great heats after the summer solstice generally end in thunderstorms; but if these do not come, in wind and rain, which last for many days.—BACON.

Early and late.

Thunder and lightning early in winter or late in fall indicate warm weather.

Heat.

Lightning brings heat.

Winter thunder.

Winter thunder,
To old folks death, to young folks plunder.

Winter's thunder,
Summer's wonder.

Thunder-storms.

A thunderstorm comes up against the wind.

Thunderstorms almost always occur when the weather is hot for the season. They are generally caused by a cold wind coming over a place where the air is much heated. They do not cool the air: it is the wind that brings them which makes the weather cooler. If a thunderstorm comes up from the east, the weather will not be cooler after it. This will not happen till another storm comes up from the west. Thunderstorms are the more violent the greater the difference of temperature between the two currents of wind which produce them.

[Thunder and lightning.]

Silence before a thunder-storm.

The air useth to be extreme hot before thunders.—BACON.

We often see, against some storm,
A silence in the heavens, the rack stand still,
The bold wind speechless, and the orb below
As hush as death: anon the dreadful thunder
Doth rend the region.—SHAKESPEARE'S "HAMLET."

Morning thunder.

When it thunders in the morning, it will rain before night.

Thunder in the morning denotes winds; at noon, showers.
BACON.

Times.

Thunder in ye morning signifies wynde, about noone rayne, in ye evening great tempest.—DIGGES.

Evening thunder.

If there be thunder in the evening, there will be much rain and showery weather.

Lightning south-east.

If in a clear and starry night it lighten in the south-east, it foretelleth great store of wind and rain to come from those parts.—HUSBANDMAN'S PRACTICE.

Sheet.

If there be sheet lightning with a clear sky on spring, summer, and autumn evenings, expect heavy rains.

Forked.

Forked lightning at night,
The next day clear and bright.

Summer.

Lightning in summer indicates good, healthy weather.

Night and morning.

Sheet lightning, without thunder, during the night, having a whitish colour, announces unsettled weather. In the west of Scotland morning lightning is regarded as an omen of bad weather.—SCOTLAND.

Distant.

Lightning without thunder after a clear day, there will be a continuance of fair weather.

Lightning in a clear sky signifies the approach of wind and rain from the quarter where it lightens; but if it lightens in different parts of the sky, there will be severe and dreadful storms.—BACON.

The distant thunder speaks of coming rain.

Disappearing.

If it sinks from the north,
It will double its wrath.
If it sinks from the south,
It will open its mouth.
If it sinks from the west,
It is never at rest.
If it sinks from the east,
It will leave us in peace.—KENT.

[Thunder and lightning.] Direction.

If the lightning is in the colder quarters of the heaven, as the north and north-east, hailstorms will follow; but if in the warmer, as the south and west, there will be showers with a sultry temperature.—BACON.

North.

Lightning under north star will bring rain in three days.

Lightning in the north will be followed by rain in twenty-four hours.

Lightning in the north in summer is a sign of heat.

North-west.

When it lightens only from the north-west, look for rain the next day.—WILLSFORD.

Thunderstorm from north-west is followed by fine, bracing weather; but thunder and lightning from north-east indicates sultry, unsettled weather.—OBSERVER AT SANTA FÉ.

East.

If the first thunder is from the east, the winter is over.
ZUÑI INDIANS.

South.

Lightning in the south, low on the horizon, indicates dry weather.—KANSAS.

South and north.

Thunder from the south or south-east indicates foul weather; from the north or north-west, fair weather.

A thunderstorm from the south is said to be followed by warmth, and from the north by cold. When the storm disappears in the east, it is a sign of fine weather.
SCOTLAND.

South or west.

If from the south or the west it lightens, expect both wind and rain from these parts.—WILLSFORD.

Rain.

After the clap there follows a heavy and abundant shower of rain.—LUCRETIUS, C. W. EMPSON'S TRANSLATION.

After much thunder, much rain.—FRANCE.

Rain and wind increase after a thunderclap.
VIRGIL'S "GEORGICS," BOOK I.

Souring milk.

Abundance depends on sour milk.
[The meaning of this is that thunderstorms aid crops.]

Increasing atmospheric electricity oxidises ammonia in the air, and forms nitric acid, which affects milk, thus accounting for the souring of milk by thunder.—MAJOR DUNWOODY.

Superstition.

When it thunders, they say the prophet (Elias) is driving in his chariot in pursuit of demons.—T. BENT (GREECE).

Lightning colours.

When the flashes of lightning appear very pale, it argues the air to be full of waterish meteors; and if red and fiery, inclining to winds and tempests.

[Thunder and lightning.]

Description.

As when two black clouds,
With heaven's artillery fraught, come rattling on
Over the Caspian; then stand, front to front
Hovering a while, till winds the signal blow
To join their dark encounter in mid air.—MILTON.

Bells. The sound of bells is supposed to dissipate thunder and lightning.—BACON.

[Church bells are still rung in the Austrian Tyrol with this object.]

Rolling. Rolling thunder which seems to be passing on foretells wind; but sharp and interrupted cracks denote storms both of wind and rain.—BACON.

Continuous. When the thunder is more continuous than the lightning, there will be great winds; but if it lightens frequently between the thunderclaps, there will be heavy showers with large drops.—BACON.

Barometer.

BAROMETER.

Variations. The variations of the barometer depend on the variations of the wind. It is highest during frost, with a north-east wind; and lowest during a thaw, with a south or south-west wind.

JENYNS.

Falling with east wind. A steady and considerable fall in the mercury during an east wind denotes that the wind will soon go round to the south, unless a heavy fall of snow or rain immediately follow: in this case, the upper clouds usually come up from the south.

BELVILLE.

Falling with north or west wind. If the mercury fall with the wind at the west, north-west, or north, a great reduction of temperature will follow: in the winter severe frosts; in the summer cold rains.—BELVILLE.

Falling with westerly wind. If the mercury fall during a high wind from the south-west, south-south-west, or west-south-west, an increasing storm is probable; if the fall be rapid, the wind will be violent, but of short duration; if the fall be slow, the wind will be less violent, but of longer continuance.—BELVILLE.

Falling with south wind. A fall of the mercury with a south wind is invariably followed by rain in greater or less quantities.—BELVILLE.

Rise and fall. Neither a sudden rise nor a sudden fall of the barometer is followed by any lasting change of weather. If the mercury rise and fall by turns, it is indicative of unsettled weather.

JENYNS.

Falling. The barometer falls for southerly and westerly winds, and for damper, stormier, and warmer weather.

[Barometer.] Rising.

A sudden rise in the barometer is very nearly as dangerous as a sudden fall, because it shows that the level is unsteady. In an ordinary gale the wind often blows hardest when the barometer is just beginning to rise, directly after having been very low.

Wind.

The barometer rises for northerly or easterly winds, and for dryer, calmer, and colder weather.

Rising in wet weather.

In wet weather, when the barometer rises much and high, and so continues for two or three days before wet weather is quite over, you may expect a continuance of fair weather for several days.—C. L. PRINCE.

Falling in fair weather.

In fair weather, when the barometer falls much and low, and thus continues for two or three days before the rain comes, you may expect much rain, and probably high winds.

C. L. PRINCE.

Rise after rain.

A sudden and considerable rise of the barometer after several hours of heavy rain, accompanied by a drying westerly wind, indicates more rain within thirty hours, and a considerable fall of the barometer.—C. L. PRINCE.

Low.

Should the barometer continue low when the sky becomes low after heavy rain, expect more rain within twenty-four hours.—C. L. PRINCE.

Wind.

When, after a succession of gales and great fluctuations of the barometer, a gale comes on from south-west, which does not cause much, if any, depression of the instrument, you may consider that more settled weather is near at hand.

C. L. PRINCE.

Fall in fine weather.

If the barometer fall gradually for several days during the continuance of fine weather, much wet will probably ensue in the end. In like manner, if it keep rising while the wet continues, the weather, after a day or two, is likely to set in fair for some time.—JENYNS.

Steady after storm.

If after a storm of wind and rain the mercury remain steady at the point to which it had fallen, serene weather may follow without a change of wind; but on the rising of the mercury rain and a change of wind may be expected.—BELVILLE.

Indication.

The height of the barometer must be above the mean corresponding to the particular wind blowing at the time to allow of weather in which any confidence can be placed.—JENYNS.

Low in fine weather.

A very low barometer is usually attendant upon stormy weather, with wind and rain at intervals, but the latter not necessarily in any great quantity. If the weather, notwithstanding a very low barometer, is fine and calm, it is not to be depended upon: a change may come on very suddenly.

JENYNS.

[Barometer.] Frost. If it freezes, and the barometer falls two or three-tenths of an inch, expect a thaw.

High with warmer weather. If the weather gets warmer while the barometer is high and the wind north-easterly, we may look for a sudden shift of wind to the south. On the other hand, if the weather becomes colder while the wind is south-westerly and the barometer low, we may look for a sudden squall or a severe storm from the north-west, with a fall of snow if it be winter-time.

Wet after a fall. When wet weather happens soon after the falling of the barometer, expect but little of it; and, on the contrary, expect but little fair weather when it proves fine shortly after the barometer has risen.—C. L. PRINCE.

Rising with warmth. During summer, if pressure and temperature increase together, expect several fine days; and if small patches of cirro-cumulus cloud should appear at a great elevation, the rise of temperature will be considerable.—C. L. PRINCE.

> If barometer and thermometer both rise together,
> It is a very sure sign of coming fine weather.

Rising after heavy rains. After heavy rains from south-west, if the barometer rises upon the wind shifting to the north-west, expect three or four fine days.—C. L. PRINCE.

Falling quickly. If the barometer falls two or three-tenths of an inch in four hours, expect a gale of wind.—C. L. PRINCE.

Oscillating. If you observe that the surface of the mercury in the cistern of the barometer vibrates upon the approach of a storm, you may expect the gale to be severe.—C. L. PRINCE.

Snow. The barometer seldom falls for snow.—C. L. PRINCE.

Summer. In summer, when the barometer falls suddenly, expect a thunderstorm; and if it does not rise again when the storm ceases, there will be several days' unsettled weather.

C. L. PRINCE.

Local. A summer thunderstorm, which does not much depress the barometer, will be very local and of slight consequence.

C. L. PRINCE.

Rising with dry weather. When the barometer rises considerably, and the ground becomes dry, although the sky remains overcast, expect fair weather for a few days. The reverse may be expected if water is observed to stand in shallow places, notwithstanding the barometer may read upwards of thirty inches.

C. L. PRINCE.

Falling without change. When the barometer falls considerably without any particular change of weather, you may be certain that a violent storm is raging at a distance.—C. L. PRINCE.

[Barometer.]

Winter. During winter, heavy rain is indicated by a decrease of pressure and an increase of temperature.—C. L. PRINCE.

Indicating frost. In winter the rising barometer indicates frost when the wind is east-north-east; and should the frost and increase of pressure continue, expect snow.—C. L. PRINCE.

Rapid fall. The barometer falls lower for high winds than for heavy rains. If the fall amount to one inch in twenty-four hours, expect a very severe gale.—C. L. PRINCE.

High and steady. A high and steady barometer is indicative of settled weather.
JENYNS.

Change. In general the barometer falls before rain; and all appearances being the same, the higher the barometer, the greater the probability of fair weather.—DALTON.

Summary. An excellent summary of the barometer rules, which are too numerous to quote here, was given by G. F. Chambers, F.R.A.S., in *The Churchman* newspaper, February 1868.

High in north. When the barometer is higher at Brest than at Nairn, while it is of about the same value at Valentia and Yarmouth, being gradually less from south to north, then the winds over Britain are *westerly*.—R. STRACHAN.

Equal east and west. When the barometer at Nairn is higher than at Brest, while its readings at Valentia and Yarmouth are about equal, the winds over Britain are *easterly*.—R. STRACHAN.

High in west. When the barometer at Valentia is higher than at Yarmouth, while its readings at Brest and Nairn are about equal, the winds over Britain are *northerly*.—R. STRACHAN.

High in east. When the barometer at Yarmouth is higher than at Valentia, while there is equality of pressure at Nairn and Brest, the winds over Britain are *southerly*.—R. STRACHAN.

Equal readings. When the barometer readings at Brest, Valentia, Nairn, and Yarmouth are nearly equal, then the winds over the British Isles are variable in direction and light in force.
MR. R. STRACHAN'S RULES.

Wind.
When the glass falls low,
Prepare for a blow;
When it rises high,
Let all your kites fly.—NAUTICAL.

First rise.
First rise after low
Foretells stronger blow.

Long notice.
Long foretold,* long last;
Short notice, soon past.—FITZROY.

* By the falling of the mercury.

BAROMETER WARNINGS.	INDICATING.
If mercury falls during a high wind from S.W., S.S.W., W., or S.	Increasing storm.
If the fall be rapid	Violent, but short.
If the fall be slow	Less violent, but longer continuance.
If the mercury falls suddenly whilst the wind is due W.	A violent storm from N.W. or N.
If the mercury, having been at its usual height (29·95), is steady or rising, while the thermometer falls and the air becomes dryer	N.W., N., or N.E. winds, or less wind, or less rain, or less snow.
If the mercury falls, while the thermometer rises and the air becomes damp	Wind and rain from S.E., S., or S.W.
When the mercury falls suddenly with a W. wind	A violent storm from N.W., N., or N.E.
If the mercury falls when the thermometer is low	Snow.
When the mercury rises, after having been some time below its average height	Less wind, or a change to N., or less wet.
With the first rise of the mercury after it has been very low (say 29 in.)	Strong wind or heavy squalls from N.W., N., or N.E.
When a gradual, continuous rise of the mercury occurs with a falling thermometer	Improved weather.
If the mercury suddenly rising, the thermometer also rises	Winds from S. or S.W.
Soon after the first rise of the mercury from a very low point	Heavy gales from N.
With a rapid rise of the mercury	Unsettled weather.
With a slow rise of the mercury	Settled weather.
With a continued steadiness of the mercury with dry air	Very fine weather.
With a rapid and considerable fall of the mercury	Stormy weather with rain, or snow.
With an alternate rising and falling of the mercury	Threatening, unsettled weather.
When the mercury falls considerably, if the thermometer be low (for the season), the wind will be N.; if high, from S.	Much wind, rain, hail, or snow, with or without lightning.
When the mercury is low, the storm being beyond the horizon	Lightning only.

FROM "WEATHER WARNINGS," BY "THE CLERK" HIMSELF, 1877.

Barometer in Ireland.

Satirical.

Very high and rising fast,
Steady rain and sure to last.
Steady high after low,
Floods of rain, or hail, or snow.
Falling fast,
Fine at last.
Rapid fall after high,
Sun at last, and very dry.

[Satirical rhyme suggested by six weeks of rain, with generally high and steady barometer.]

"SYMONS' METEOROLOGICAL MAGAZINE," OCTOBER 1892.

Thermometer.

THERMOMETER.

Temperature increasing.

If the temperature increases between 9 p.m. and midnight, when the sky is cloudless, expect rain; and if during a long and severe frost the temperature increases between midnight and sunrise, expect a thaw.—C. L. PRINCE.

Different levels.

The greater the difference between the lowest temperature of the air at four feet from the ground, and that of terrestrial radiation under a cloudless sky, the less will be the probability of the existing state of weather continuing, and *vice versâ*.—C. L. PRINCE.

Hygrometer.

HYGROMETER.

Indications.

The greater the difference between the readings of the wet and dry-bulb thermometers, the greater will be the probability of fine weather, and *vice versâ*.—C. L. PRINCE.

Telescope.

TELESCOPE.

Air tremor.

If the images of stars or the moon appear ill defined and surrounded by much atmospheric tremor, expect both wind and rain. The greater the tremor, the sooner the change, except when the wind is easterly.—C. L. PRINCE.

Animals.

The observations of naturalists, shepherds, herdsmen, and others who have been brought much into contact with animals, have proved most clearly that these creatures are cognisant of approaching changes in the state of the air long before we know of their coming by other signs. To many kinds of animals, birds, and insects, the weather is of so much more importance than to us, that it would be wonderful if nature had not provided them with a more keenly prophetic instinct in this respect. The occurrence of a storm would, doubtless, be the means of depriving some of the Carnivora of a meal, and it is known that utter destruction would occur to the nests of some birds if the tenants were absent during a gale of wind or a pelting shower; while to vast numbers of insects the state of the weather for the fraction of a week may determine the whole time during which they can enjoy their little lives. To enable all these creatures to prepare for coming trouble, they seem to have been fitted with what is to us an unknown sense informing them of minute changes in the atmosphere, and it has long been observed that they eat with more avidity, return to their homes, or become unusually restless before the coming of the danger of which they are forewarned.

This is a subject on which there is still a great deal to be learnt, and I hope naturalists will continue to collect notes on so important a matter.

ANIMALS.

Seeking cover. When animals seek sheltered places instead of spreading over their usual range, an unfavourable change is probable.

Crowding. If animals crowd together, rain will follow.

[DOGS.] The unusual howling of dogs portends a storm.

Uneasy. Dogs making holes in the ground, howling when any one goes out, eating grass in the morning, or refusing meat, are said to indicate coming rain.—MAJOR DUNWOODY.

Eating grass. When dogs eat grass, it will be rainy.

Rolling. If dogs roll on the ground and scratch, or become drowsy and stupid, it is a sign of rain.

Sleeping. If spaniels sleep more than usual, it foretells wet weather.

Stretched out.

Sign, too, of rain: his outstretched feet the hound
Extends, and curves his belly to the ground.

J. LAMB'S "ARATUS."

[CATS.]

Sneezing. When a cat sneezes, it is a sign of rain.

Movements. Cats are observed to scratch the wall or a post before wind, and to wash their faces before a thaw; they sit with their backs to the fire before snow.—SCOTLAND.

Licking. They say here that if a cat licks herself with her face turned towards the north, the wind will soon blow from that dangerous quarter.—T. BENT (GREECE).

Imprisoned. It is an Irish saying that putting the cat under the pot will bring bad weather, and this is sometimes done in jest to prevent a guest from departing.—FOLK-LORE JOURNAL.

Enraged. Cats with their tails up and hair apparently electrified indicate approaching wind,—or a dog.

Washing. The cardinal point to which a cat turns and washes her face after a rain shows the direction from which the wind will blow.

The old woman promised him a fine day to-morrow because the cat's skin looked bright. [*Animals.*] *Cat's skin.*

Sailors dislike to see the cat on board ship unusually playful or quarrelsome, and they say the cat has a gale of wind in her tail.—BRAND. *Playful.*

When the cat scratches the table legs, a change is coming. *Scratching.*

> While rain depends, the pensive cat gives o'er
> Her frolics, and pursues her tail no more.—BROOME. *Pensive.*

When cats wipe their jaws with their feet, it is a sign of rain, and especially when they put their paws over their ears in wiping. *Wiping jaws.*

If horses stretch out their necks and sniff the air, rain will ensue. [HORSES.] *Sniffing.*

Horses sweating in the stable is a sign of rain. *Sweating.*

If they start more than ordinary and are restless and uneasy, or if they assemble in the corner of a field with heads to leeward, expect rain. *Restless.*

If young horses do rub their backs against the ground, it is a sign of great drops of rain to follow.
HUSBANDMAN'S PRACTICE. *Rolling.*

Horses and mules, if very lively without apparent cause, indicate cold. *Lively.*

In the notes to *Quentin Durward*, by Sir Walter Scott, there is an anecdote of Louis XI., who, refusing to believe the weather prophecy of a charcoal burner, got soaked with rain. When the man was asked how he was able so well to predict the weather, he replied that his own donkey was his prophet, and on the approach of rain pricked his ears forward, walked slowly, and tried to rub himself against walls. [ASSES.] *Restless.*

If asses hang their ears downward and forward, and rub against walls, rain is approaching. *Rubbing.*

If asses bray more frequently than usual, it foreshows rain. *Braying.*

> Hark! I hear the asses bray;
> We shall have some rain to day.—RUTLAND.

> It is time to stack your hay and corn
> When the old donkey blows his horn.

When cattle lie down during light rain, it will soon pass.
WILTSHIRE. [CATTLE.] *Lying down.*

[Animals.] Cattle on hills. When cattle remain on hilltops, fine weather to come.

DERBYSHIRE.

[Cows.] When cows fail their milk, expect stormy and cold weather.

When cows bellow in the evening, expect snow that night.

Various signs. There are other sayings about cows—such as, if they stop and shake their feet, or refuse to go to pasture in the morning, or when they low and gaze at the sky, or lick their forefeet, or lie on the right side, or rub themselves against posts, or lie down early in the day, it indicates rain to come.

MAJOR DUNWOODY.

The cattle also concerning the vapour.—JOB xxxvi. 33.

In autumn.

When autumn's days are nearly passed away,
And winter hastens to assume his sway,
Mark if the kine and sheep at eventide
Toss up their hornèd heads, with nostril wide
Imbibe the northern breeze, and furious beat
The echoing meadows with their cloven feet;
For tyrant winter comes with icy hand,
Heaping his snowy ridges on the land,
Blasting Pomona's hopes with shrivelling frost,
While Ceres mourns her golden treasure lost.

J. LAMB'S "ARATUS."

Lying down. When kine and horses lie with their heads upon the ground, it is a sign of rain.

Cow.

When a cow tries to scratch its ear,
It means a shower is very near;
When it thumps its ribs with its tail,
Look out for thunder, lightning, hail.

NEW JERSEY, U.S.

[BULLS.] If bulls lick their hoofs or kick about, expect much rain.

Leading. If the bull lead the van in going to pasture, rain must be expected; but if he is careless, and allow the cows to precede him, the weather will be uncertain.

[OXEN.] If oxen be seen to lie along upon the left side, it is a token of fair weather.—HUSBANDMAN'S PRACTICE.

Licking. When oxen do lick themselves against the hair, it betokeneth rain to follow shortly after.—HUSBANDMAN'S PRACTICE.

Sniffing. If oxen turn up their nostrils and sniff the air, or if they lick their forefeet, or lie on their right side, it will rain.

Turning tail to wind. "When that white stirk o' ours turns her tail to the wind, you're sure to ha'e rain in half an hour."

COWBOY TO "OLD MOORE," THE ALMANACK MAKER.

He taught us erst the heifer's tail to view;
When stuck aloft, that showers would straight ensue.

GAY.

[*Animals.*]
Bullocks.

The herdsmen too, while yet the skies are fair,
Warned by their bullocks, for the storm prepare—
When with rough tongue they lick their polished hoof,
When bellowing loud they seek the sheltering roof,
When from the yoke at close of day released,
On his right side recumbs the wearied beast:
When keenly pluck the goats the oaken bough;
And deeply wallows in the mire the sow.

J. LAMB'S "ARATUS."

Goats.

The goat will utter her peculiar cry before rain.

Goats leave the high grounds and seek shelter before a storm.

SCOTLAND.

Goats and sheep.

If goats and sheep quit their pastures with reluctance, it will rain the next day.

[SHEEP.]

If old sheep turn their backs towards the wind, and remain so for some time, wet and windy weather is coming.

Frisky.

All shepherds agree in saying that before a storm comes sheep become frisky, leap, and butt or "box" each other.

FOLK-LORE JOURNAL.

If sheep gambol and fight, or retire to shelter, it presages a change in the weather.

The shepherd, as afield his charge he drives,
From his own flock prognostics oft derives.
When they impetuous seek the grassy plain,
He marks the advent of some storm or rain;
And when grave rams and lambkins full of play
Butt at each other's heads in mimic fray;
When the horned leaders stamp the dusty ground
With their forefeet—all fours the young ones bound;
When homeward, as the shades of night descend,
Reluctantly and slow their way they wend,
Stray from the flock, and linger one by one,
Heedless of shepherd's voice and missive stone.

Returning slowly.

J. LAMB'S "ARATUS."

Feeding.

Old sheep are said to eat greedily before a storm, and sparingly before a thaw. When they leave the high grounds, and bleat much in the evening and during the night, severe weather is expected. In winter, when they feed down the hill, a snowstorm is looked for; when they feed up the burn, wet weather is near.

[Animals.] *Sheep.*

If sheep feed up-hill in the morning, sign of fine weather.
DERBYSHIRE.

When sheep turn their backs to the wind, it is a sign of rain.

[PIGS.] *Restless.*

Swine are so terrified and disturbed and discomposed when the wind is getting up, that countrymen say that this animal alone sees the wind, and that it must be frightful to look at.
BACON.

Carrying straw.

Hogs crying and running unquietly up and down with hay or litter in their mouths foreshadows a storm to be near at hand.—THOMAS WILLSFORD.

When pigs carry straw to their sties, bad weather may be expected.

Grumphie smells the weather,
An' grumphie sees the wun';
He kens when clouds will gather,
An' smoor the blinkin' sun;
Wi' his mou' fu' o' strae
He to his den will gae;
Grumphie is a prophet,—
Wat weather we will hae.—GALLOWAY.

When pigs carry sticks,
The clouds will play tricks;
Wallowing.
When they lie in the mud,
No fears of a flood.

Seeing wind.

Pigs can see the wind.

Rubbing.

Hogs rubbing themselves in winter indicates an approaching thaw.—MAJOR DUNWOODY.

Wolf.

When through the dismal night the lone wolf howls,
Or when at eve around the house he prowls,
And, grown familiar, seeks to make his bed,
Careless of man, in some outlying shed,—
Then mark: ere thrice Aurora shall arise,
A horrid storm will sweep the blackened skies.
J. LAMB'S "ARATUS."

Beaver.

In early and long winters the beaver cuts his winter supply of wood, and prepares his house one month earlier than in mild, late winters.—MAJOR DUNWOODY.

Rats.

If rats are more restless than usual, rain is at hand.

Mice.

E'en mice ofttimes prophetic are of rain,
Nor did our sires their auguries disdain,
When loudly piping with their voices shrill,
They frolicked, dancing on the downy hill.
J. LAMB'S "ARATUS."

If mice run about more than usual, wet weather may be expected.

[*Animals.*] *Mice.*

Moles plying their works, in undermining the earth, foreshows rain; but if they do forsake their trenches and creep above ground in summer-time, it is a sign of hot weather; but when on a sudden they do forsake the valleys and low grounds, it foreshows a flood near at hand; but their coming into meadows presages fair weather, and for certain no floods.

THOMAS WILLSFORD.

[MOLES.] *Busy.*

Previous to the setting in of winter the mole prepares a sort of basin, forming it in a bed of clay, which will hold about a quart. In this basin a great quantity of worms is deposited; and, in order to prevent their escape, they are partly mutilated, but not so much as to kill them. On these worms the moles feed in the winter months. When these basins are few in number, the following winter will be mild.

GARDENER'S CHRONICLE.

Storing food.

[I have asked several mole-catchers in Hampshire (near Southampton, where moles are very numerous) whether the above was true, and they all answered in the affirmative.—C. W. EMPSON.]

When the mole throws up fresh earth during a frost, it will thaw in less than forty-eight hours.

Digging.

If moles throw up more earth than usual, rain is indicated.

Hares take to the open country before a snowstorm.

SCOTLAND.

Hares.

When squirrels lay in a large supply of nuts, expect a cold winter; but—

When he eats them on the tree,
Weather as warm as warm can be.

Squirrels.

If weasels and stoats are seen running about much in the forenoon, it foretells rain in the after-part of the day.—SCOTLAND.

Weasels, stoats, etc.

It will rain if bats cry much or fly into the house.

[BATS.]

If bats abound and are vivacious, fine weather may be expected.

Numerous.

Observe which way the hedgehog builds her nest,
To front the north or south, or east or west;
For if 'tis true that common people say,
The wind will blow the quite contrary way.
If by some secret art the hedgehog knows,
So long before, which way the winds will blow,
She has an art which many a person lacks
That thinks himself fit to make our almanacks.

POOR ROBIN'S ALMANACK, 1733.

[HEDGEHOGS.]

[Animals.] Hedgehogs. Hiding. Hedgehogs conceal themselves in their holes before a change of wind from north-east to south.—PLINY.

As hedgehogs doe foresee ensuing stormes.
BODENHAM'S "BELVEDERE," 1600.

Burrows. The hedgehog commonly hath two holes or vents in his den or cave, the one towards the south and the other towards the north; and look which of them he stops—thence will great storms and winds follow.—HUSBANDMAN'S PRACTICE.

Birds.

BIRDS. *Departing.*

When numerous birds their island home forsake,
And to firm land their airy voyage make,
The ploughman, watching their ill-omened flight,
Fears for his golden fields a withering blight.
Not so the goatherd—he their advent hails,
As certain promise of o'erflowing pails.
J. LAMB'S "ARATUS."

Silent. If the birds be silent, expect thunder.

Returning. If birds that dwell in trees return eagerly to their nests, and leave their feeding-ground early, it is a sign of storms; but when a heron stands melancholy on the sand, or a raven stalks about, it only denotes rain.—BACON.

If birds return slowly to their nests, rain will follow.

Flight. Migratory birds fly south from cold, and north from warm weather. When a severe cyclone is near, they become puzzled and fly in circles, dart in the air, and can be easily decoyed.—NORTH CAROLINA.

Restless. When birds of long flight—rooks, swallows, or others—hang about home, and fly up and down or low, rain or wind may be expected.

[SMALL BIRDS.] *Washing.* If small birds seem to duck and wash in the sand, it is held to be a sign of coming rain.

Summer. When summer birds take their flight, summer goes with them.

Arriving. When the fieldfare, redwing, starling, swan, snowfleck, and other birds of passage arrive soon from the north, it indicates the probability of an early and severe winter.—SCOTLAND.

[FOWLS.] If the fowls huddle together outside the henhouse instead of going to roost, there will be wet weather.
FOLK-LORE JOURNAL.

If fowls grub in the dust and clap their wings, or if their wings droop, or if they crowd into a house, it indicates rain. [*Birds.*] *Fowls.*

If fowls roll in the sand,
Rain is at hand. *Rolling.*

When they look towards the sky, or roost in the daytime, expect rain; but if they dress their feathers during a storm, it is about to cease; while their standing on one leg is considered a sign of cold weather. When fowls collect together, and pick or straighten their feathers, expect a change. *Restless.*

Fowls will run to shelter and stay there if they think the weather will clear; but if they see it is to be wet all day, they come out and face it.—UNITED STATES.

If the cock moult before the hen,
We shall have weather thick and thin;
But if the hen moult before the cock,
We shall have weather hard as a block. *Moulting.*

If the cock drink in summer, it will rain a little after.—ITALY. *Drinking.*

If cocks crow late and early, clapping their wings unusually, rain is expected. *Crowing.*

If the cock goes crowing to bed,
He'll certainly rise with a watery head.

If ducks or drakes do shake and flutter their wings when they rise, it is a sign of ensuing water. [DUCKS.] *Restless.*

HUSBANDMAN'S PRACTICE.

When ducks are driving through the burn,
That night the weather takes a turn. *Driving.*

Divers and ducks prune their feathers before a wind; but geese seem to call down the rain with their importunate cackling.—BACON. [DUCKS AND GEESE.]

If ducks and geese fly backwards and forwards, and continually plunge in water and wash themselves incessantly, wet weather will ensue. *Uneasy.*

If the wild geese gang out to sea,
Good weather there will surely be. [GEESE.]

Wild geese, wild geese, ganging to the sea,
Good weather it will be; *Flying seawards.*
Wild geese, wild geese, ganging to the hill,
The weather it will spill.—MORAYSHIRE. *Flying inland.*

Wild geese moving south indicates approaching cold weather; moving north indicates that most of the winter is over. *Flying south.*

UNITED STATES.

[Birds.] Geese. When wild geese fly to the south-east in the fall, in Kansas, expect a blizzard.—UNITED STATES.

Flying directly south and very high indicates a cold winter.

Flying low. When flying low and remaining along the river, they indicate a warm winter in Idaho. For spring, just the reverse when flying north.—"OLD SETTLER," UNITED STATES.

Passing lakes. Wild geese flying past large bodies of water indicate change of weather; going south, cold; going north, warm.
UNITED STATES.

Flight. It is said that the flight of wild geese is always either in the form of letters or of figures, and that the figures denote the number of weeks of frost that would follow their appearance.

Breast-bone. When the goose-bone, exposed to air, turns blue, it indicates rain; when it retains its colour, expect clear weather.

The whiteness of a goose's breast-bone is superstitiously thought to indicate or foreshow the amount of snow during winter.

If the November goose-bone be thick,
So will the winter weather be;
If the November goose-bone be thin,
So will the winter weather be.

[TURKEYS.] Turkeys perched on trees and refusing to descend indicates snow.

Flight. Water turkeys flying against the wind indicates falling weather.
UNITED STATES.

[GUINEA-FOWL.] This bird, called the "come-back" in Norfolk, is regarded as an invoker of rain. It often continues clamorous throughout the whole of rainy days.—C. SWAINSON.

Rain. Guinea-fowls squall more than usual before rain.

[SWANS.] When swans fly, it is a sign of rough weather.
J. W. G. GUTCH.

Flight. If the swan flies against the wind, it is a certain indication of a hurricane within twenty-four hours, generally within twelve.
CORRESPONDENT IN THE "ATHENÆUM," VOL. III., P. 229.

Nest. The swan is said to build its nest high before floods, but low when there will not be unusual rains.

Orkneys. When the white swan visits the Orkneys, expect a continued severe winter.—SCOTLAND.

[PARROTS.] Clamorous as a parrot against rain.—SHAKESPEARE.

Rain. Parrots whistling indicate rain.

[Birds.]
Parrots and canaries.

It is said that parrots and canaries dress their feathers and are wakeful the evening before a storm.

Macaws.

The feathers of the blue macaw turn a greenish hue before rain.—DR. THORNTON.

[PEACOCKS.]

When the peacock loudly bawls,
Soon we'll have both rain and squalls.

Crying.

If peacocks cry in the night, there is rain to fall.

Rain.

The strutting peacock yawling 'gainst the rain.—DRAYTON.

When the peacock's distant voice you hear,
Are you in want of rain? Rejoice, 'tis almost here.

Pea-fowl.

Pea-fowl utter loud cries before a storm, and select a low perch.

[PIGEONS.]

Pigeons wash before rain.—J. W. G. GUTCH.

Returning.

Doves or pigeons coming later home to the dove-house in the evening than ordinary, it is a token of rain.
HUSBANDMAN'S PRACTICE.

If pigeons return home slowly, the weather will be wet.

Partridge.

If the partridge sings when the rainbow spans the sky,
There is no better sign of wet than when it isn't dry.
SPANISH RHYME.

Ptarmigans.

The frequently repeated cry of the ptarmigan low down on the mountains during frost and snow indicates more snow and continued cold.—SCOTLAND.

Woodcocks.

An early appearance of the woodcock indicates the approach of a severe winter.—UNITED STATES.

Grouse.

The gathering of grouse into large flocks indicates snow. Their approach to the farmyard is a sign of severe weather—frost and snow. When they sit on dykes on the moor, rain only is expected.—SCOTLAND.

Quails.

When quails are heard in the evening, expect fair weather next day.

Quails are more abundant during an easterly wind.
UNITED STATES.

Snipe.

The drumming of the snipe in the air, and the call of the partridge, indicate dry weather and frost at night to the shepherds of Garrow.—SCOTLAND.

[PRAIRIE CHICKENS.]

Prairie chickens coming into the creeks and timber indicates cold weather.

Cold.

When the prairie chicken sits on the ground with all its feathers ruffled, expect cold weather.—UNITED STATES.

Rooks.

When rooks seem to drop in their flight, as if pierced by a shot, it is considered to foretell rain.

[Birds.] Rooks.

This "tumbling" of rooks is amongst the best-known signs of rain in places where those birds are found.

The low flight of rooks indicates rain. If they feed busily, and hurry over the ground in one direction, and in a compact body, a storm will soon follow. When they sit in rows on dykes and palings, wind is looked for. When going home to roost, if they fly high, the next day will be fair, and *vice versâ.* If when flying high they dart down and wheel about in circles, wind is foreshown. In autumn and winter, if after feeding in the morning they return to the rookery and hang about it, rain is to be expected.—SCOTLAND.

When rooks fly sporting high in air,
It shows that windy storms are near.

If rooks stay at home, or return in the middle of the day, it will rain; if they go far abroad, it will be fine.

DEVONSHIRE.

Crows.

When crows go to the water, if they beat it with their wings, throw it over them, and scream, it foreshows storms.—BACON.

Thrush.

The missel-thrush (in Hampshire called the "storm-cock") sings particularly loud and long before rain.

When the thrush sings at sunset, a fair day will follow.

Blackbirds.

When the voices of blackbirds are unusually shrill, or when blackbirds sing much in the morning, rain will follow.

[CUCKOO.] *Early.*

Bad for the barley, and good for the corn,
When the cuckoo comes to an empty thorn.

If the cuckoo sings when the hedge is brown,
Sell thy horse and buy thy corn.—WELSH.

[You will not be able to afford horse corn.]

Late.

If the cuckoo sings when the hedge is green,
Keep thy horse and sell thy corn.

[It will be so plentiful that you will have enough and to spare.]

MISS JACKSON'S "SHROPSHIRE FOLK-LORE."

Midsummer.

If the cuckoo does not cease singing at midsummer, corn will be dear.

Rain.

Hesiod mentions the singing of a bird which he calls "kokkux" as foreboding three days' rain.—C. SWAINSON.

Cuckoos and figs.

In ancient Greece the young figs and the cuckoos came together; so the same word, "kokkux," served for both.

Gowk storms.

Spring gales about the equinox have been called "gowk storms," because they follow the cuckoo.

Low lands.

When the cuckoo is heard in low lands, it indicates rain; on high lands, fair weather.

[Birds.] Nightingale and rose.

In Asia the rose and nightingale were similarly allied.

[WOODPECKERS.]

When woodpeckers are much heard, rain will follow.

The call of the *heigh-ho* (woodpecker) forebodes rain.
SHROPSHIRE.

Leaving.

When the woodpecker leaves, expect a hard winter. When woodpeckers peck low on the trees, expect warm weather.

Pecking trees.

The ivory-billed woodpecker, commencing at the bottom end of a tree, and going to the top, removing all the outer bark, indicates a hard winter, with deep snow.—UNITED STATES.

Crying.

The yaffel, or green woodpecker (called also the "rain-bird"), cries at the approach of rain, and is described as "laughing in the sun, because the rain is coming."

Magpies.

For anglers in spring it is always unlucky to see single magpies; but two may always be regarded as a favourable omen. And the reason is, that in cold and stormy weather one magpie alone leaves the nest in search of food, the other remaining sitting with the eggs or the young ones; but when two go out together, it is only when the weather is mild and warm, and favourable for fishing.

Magpies flying three or four together and uttering harsh cries predict windy weather.

Jackdaws.

When three daws are seen on St. Peter's vane together,
Then we're sure to have bad weather.—NORWICH.

Shower-bringing daws
Shall caw their last.—TOOKE'S "LUCIAN."

Titmouse.

The titmouse foretells cold, if crying, "Pincher."

The saw-like note of the great titmouse foretells rain.
C. SWAINSON.

Ravens.

Ravens, when they croak continuously, denote wind; but if the croaking is interrupted or stifled, or at longer intervals, they show rain.—BACON.

If ravens croak three or four times and flap their wings, fine weather is expected.

Raven and rook.

The corbie said unto the craw,
"Johnnie, fling your plaid awa'";
The craw says unto the corbie,
"Johnnie, fling your plaid about ye."

[In Scotland it is believed that if the raven cries first in the morning, it will be a good day; if the rook, the reverse.—C. SWAINSON.]

[Birds.]
Screech-owl. A screeching owl indicates cold or storm.

[Owls.] If owls hoot at night, expect fair weather.

Change. The whooping of an owl was thought by the ancients to betoken a change of weather, from fair to wet, or from wet to fair. But with us an owl, when it whoops clearly and freely, generally shows fair weather, especially in winter.—Bacon.

Screaming. If owls scream during bad weather, there will be a change.

Dirt-owl. The dirt-bird (or dirt-owl) sings, and we shall have rain.

[Robins.] If robins are seen near houses, it is a sign of rain.

Singing. On a summer evening, though the weather may be in an unsettled and rainy state, he (the robin) sometimes takes his stand on the topmost twig, or on the housetop, singing cheerfully and sweetly. When this is observed, it is an unerring promise of succeeding fine days. Sometimes, though the atmosphere is dry and warm, he may be seen melancholy, chirping and brooding in a bush, or low in a hedge: this promises the reverse of his merry lay and exalted station.

Anecdotes of the Animal Kingdom, "Saturday Magazine," February 11th, 1837.

In morning. Robins indicate the approach of spring. Long and loud singing of robins in the morning denotes rain. Robins will perch on the topmost branches of trees and whistle when a storm is approaching.

In bushes.
On barns.

If the robin sings in the bush,
Then the weather will be coarse;
If the robin sings on the barn,
Then the weather will be warm.—East Anglia.

On ground. If a robin sings on a high branch of a tree, it is a sign of fine weather; but if one sings near the ground, the weather will be wet.—Oswestry.

Starlings, etc. If starlings and crows congregate together in large numbers, expect rain.

[Swallows.]
High flight.

When swallows fleet, soar high, and sport in air,
He told us that the welkin would be clear.—Gay.

Low flight. If swallows touch the water as they fly, rain approaches.

Swallows and swifts. When there are many more swifts than swallows in the spring, expect a hot and dry summer.—C. L. Prince.

[Martins.] When martins appear, winter has broken.

Frost. No killing frost after martins.

Rain. Martins fly low before and during rainy weather.

Major Dunwoody.

The plaintive note of the "shilfa" or "sheely" (chaffinch) is interpreted as a sign of rain. When, therefore, the boys hear it, they first imitate it, and then rhymingly refer to the expected consequences :— [*Birds.*] *Finches.*

Weet-weet !
Dreep-dreep ! SCOTLAND.

When the finch chirps, rain follows.

If sparrows chirp a great deal, wet weather will ensue. *Sparrows.*

If the hedge-sparrow is heard before the grape-vine is putting forth its buds, it is said that a good crop is in store. *Hedge-sparrow.*

If larks fly high and sing long, expect fine weather. [LARKS.]

Field-larks congregating in flocks indicate severe cold. *Flocks.*

When wrens are seen in winter, expect snow. *Wrens.*
UNITED STATES.

A heron, when it soars high, so as sometimes to fly above a low cloud, shows wind; but kites flying high show fair weather.—BACON. [HERONS.] *Flight.*

When the heron or bittern flies low, the air is gross and thickening into showers.

Herons in the evening flying up and down, as if doubtful where to rest, presages some evil-approaching weather. *Restless.*
THOMAS WILLSFORD.

Mark yearly when, among the clouds on high, [CRANES.] *Crying.*
Thou hear'st the shrill crane's migratory cry,
Of ploughing-time the sign and wintry rains.
HESIOD'S WORKS, ELTON'S TRANSLATION.

Their high, aerial flight the cranes suspend,
And to the earth in broken ranks descend. *Alighting.*
J. LAMB'S "ARATUS."

[A sign of bad weather.]

And when the cranes their course unbroken steer, *Flight.*
Beating with clanging wings the echoing air,
These hail, prognostics sure of weather fair.
J. LAMB'S "ARATUS."

The prudent husbandman, while autumn lasts, *Autumn.*
His precious seed on the broad furrow casts,
And fearless marks the marshalled cranes on high,
Seeking in southern climes a milder sky.
Not so the idle farmer, who delays,
And trusts to treacherous winter's shortened days. *Winter.*
He hears their screams and clanging wings with fear,
Prognostics sure of frost-bound winter near.
J. LAMB'S "ARATUS."

[Birds.] Cranes. Noisy. Cranes soaring aloft and quietly in the air foreshows fair weather; but if they make much noise, as consulting which way to go, it foreshows a storm that's near at hand.

THOMAS WILLSFORD.

Early. If cranes appear in autumn early, a severe winter is expected.

Marsh harriers. It is said in Wiltshire that the marsh harriers, or dunpickles (*Circus rufus*), alight in great numbers on the downs before rain.—C. SWAINSON.

Hawks. When men-of-war hawks fly high, it is a sign of a clear sky; When they fly low, prepare for a blow.

Kingfishers.

The peaceful kingfishers are met together
About the decks, and prophesy calm weather.—WILD.

Dotterel.

When dotterel do fast appear,
It shows that frost is very near;
But when the dotterel do go,
Then you may look for heavy snow.—SCOTLAND.

Fulmar. If the fulmar seek land, it is a sign to the inhabitants of St. Kilda that the west wind is far off.

Water-fowl. Water-fowl meeting and flocking together, but especially sea-gulls and coots flying rapidly to shore from the sea or lakes, particularly if they scream, and playing on the dry land, foreshow wind; and this is more certain if they do it in the morning.—BACON.

[SEA-BIRDS.] If sea-fowl retire to the shore or marshes, a storm approaches.

Flight. When sea-birds fly out early and far to seaward, moderate winds and fair weather may be expected. When they hang about the land or over it, sometimes flying inland, expect a strong wind with stormy weather.

[GULLS.] *Sitting on land.*

Sea-gull, sea-gull, sit on the sand;
It's never good weather while you're on the land.

SCOTLAND.

Sea-gulls in the field indicate a storm from south-east.

Arrival. The arrival of sea-gulls from the Solway Frith to Holywood, Dumfriesshire, is generally followed by a high wind and heavy rain from the south-west.

Noisy. Sea-mews early in the morning making a gaggling more than ordinary foretoken stormy and blustering weather.

Numerous. When sea-mews appear in unwonted numbers, expect rain and high south-west winds.

Petrels gathering under the stern of a ship indicate bad weather. *[Birds.] Petrels.*

The stormy petrel is found to be a sure token of stormy weather. When these birds gather in numbers in the wake of a ship, the sailors feel sure of an impending tempest. *Stormy petrel.*

In the English Channel the curlew flying on dark nights is considered as a sure precursor of an east wind. *Curlew.*

Fish, Molluscs, etc.

FISH.

Fishes rise more than usual at the approach of a storm. In some parts of England they are said not to bite so well before rain. *Rising.*

When fish bite readily and swim near the surface, rain may be expected: they become inactive just before thunder-showers. *Feeding.*

Fish bite the least
With wind in the east.

When porpoises and whales spout about ships at sea, storms may be expected. *Whales.*

[*Note.*—The whales are for the purposes of this work considered as among the fishes.—R. I.]

Porpoises are said to swim in the direction from which the wind is coming: they run into bays and round islands before a storm. [PORPOISES.]

Porpoises in harbour indicate coming storm. *In harbour.*

When porpoises swim to windward, foul weather will ensue within twelve hours. *Swimming to windward.*

Dolphins pursuing one another in calm weather foreshow wind, and from that part whence they fetch their frisks; but if they play in rough weather, it is a sign of a coming calm. [DOLPHINS.] *Playing.*

THOMAS WILLSFORD.

Dolphins, as well as porpoises, when they come about a ship and sport and gambol on the surface of the water, betoken a storm; hence they are regarded as unlucky omens by sailors.

As dolphins heave their backs above the wave,
Prognosticating angry tempests black.

DANTE'S "INFERNO," CANTO XXII., LINE 19.

Dolphins sporting in a calm sea are thought to prognosticate wind from that quarter whence they come; but if they play in a rough sea, and throw the water about, it will be fine. Most other kinds of fish, when they swim at the top of the water, or sometimes leap out of it, foretell rain —BACON.

[Fish.] Dolphins. If dolphins are seen to leap and toss, fine weather may be expected, and the wind will blow from the quarter in which they are seen.

Sharks. Sharks go out to sea at the approach of a cold wave.

[CAT-FISH.] Fish swim up stream and cat-fish jump out of the water before rain.

Skin. If the skin on the belly of the cat-fish is unusually thick, it indicates a cold winter; if not, a mild winter will follow.

NEGRO.

[COD-FISH.] *Ballast.* The cod is said to take in ballast before a storm. It is said by Sergeant McGillivray, Signal Corps, U.S.A., that there is one well-authenticated instance of this saying. A number of cod were taken twelve hours before a severe gale, and it was found that each had swallowed a number of small stones, some of the stones weighing three or four ounces.

[SALMON AND TROUT.] *Plentiful.* When salmon and trout are plentiful in the river (Columbia), it is a sign that there has been abundance of rain in the surrounding country.

Not biting.

When trout refuse bait or fly,
There ever is a storm a-nigh.

Bass. On Lake Ontario black bass leave shoal water before a thunderstorm. This has been observed twenty-four hours before the storm.

Eels. If eels are very lively, it is a sign of rain.

Mullet. Mullet run south on the approach of cold northerly wind and rain.

Pike. When pike lie on the bed of a stream quietly, expect rain or wind.

Black-fish. Black-fish in schools indicate an approaching gale.

Loach. The loach is said to be restless before stormy weather. The lake loach of the Continent (*Colitis fossilis*) remains at rest in the mud in calm weather; but when a storm approaches, it rises to the surface and moves about uneasily.

J. W. G. GUTCH.

Cockles. Cockles, it is said, have more gravel sticking to their shells before a tempest.—THOMAS WILLSFORD.

Cuttle-fish. Cuttle-fish swimming on the surface portend a storm.

THOMAS WILLSFORD.

Clam. Air bubbles over the clam-beds indicate rain.

Sea-anemone. The sea-anemone closes before rain, and opens for fine, clear weather.—J. W. G. GUTCH.

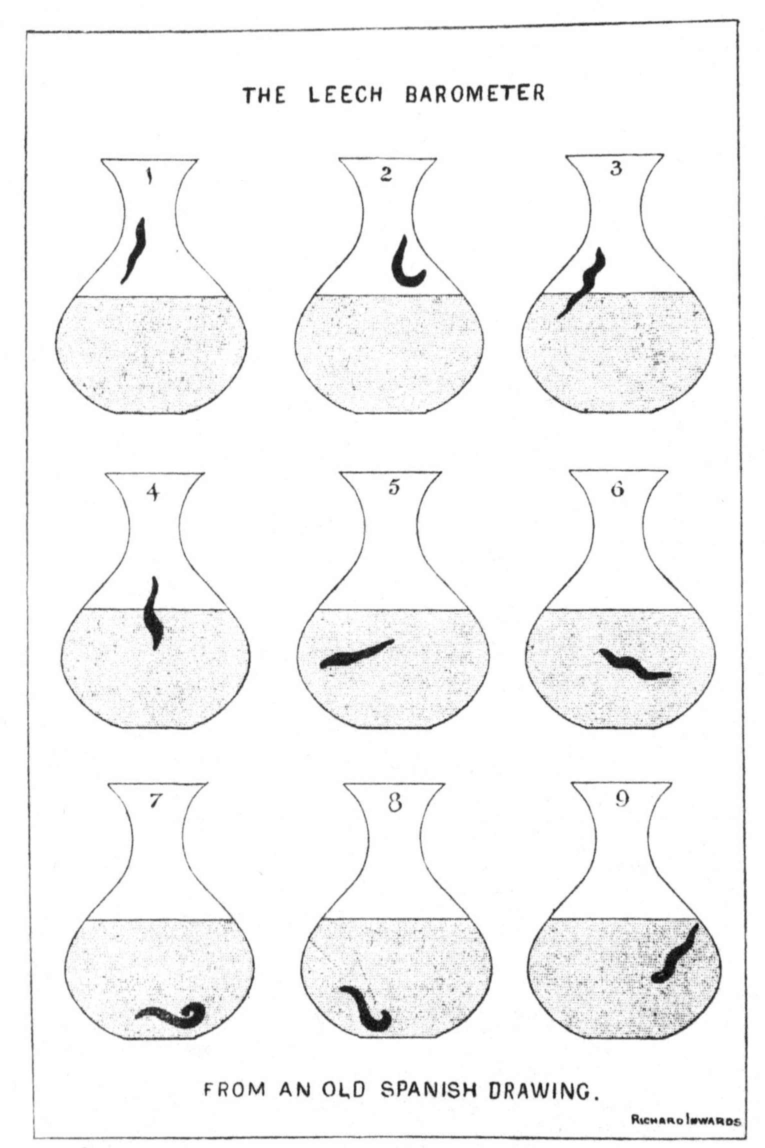
THE LEECH BAROMETER
1
2
3
4
5
6
7
8
9
FROM AN OLD SPANISH DRAWING.
RICHARD INWARDS

See page 143.

Sea-urchins striving to thrust themselves into the mud, or to cover their bodies with sand, foreshow a storm.

[SEA-URCHIN OR ECHINUS.]

THOMAS WILLSFORD.

Sinking.

The echinus is said to sink to the bottom of the sea and fasten itself firmly to sea-weeds, etc., before a storm.—E. DARWIN.

[LEECHES.]

Plate 2.

The ordinary medicinal leech has been long regarded as a weather prophet, and I met with an old Spanish drawing [see Plate 2] in Seville, giving nine positions of the leech, with nine verses describing his behaviour under various weather conditions. On the top of the drawing was the inscription, *Dios sobre todo* (God above all). The verses were to the following effect (the numbers refer to those on the drawing):—

Stationary.

1. If the leech take up a position in the bottle's neck, rain is at hand.

Curled up.

2. If he form a half-moon, when he is out of the water and sticking to the glass, sure sign of a tempest.

Restless.

3. If he is in continual movement, thunder and lightning soon.

4. If he seem as if trying to raise himself from the surface of the water, a change in the weather.

Sluggish.

5. If he move slowly close to one spot, cold weather.

Agitated.

6. If he move rapidly about, expect strong wind when he stops.

Coiled.

7. If he lie coiled up on the bottom, fine, clear weather.

8. If forming a hook, clear and cold weather.

Fixed.

9. If in a fixed position, very cold weather is certain to follow.

Tempest prognosticator.

Dr. Merryweather, of Whitby, has gone the length of contriving an apparatus by which one at least of twelve leeches confined in bottles of water rang a little bell when a tempest was expected. He showed this at the Great Exhibition of 1851, and advised the Government to establish leech-warning stations along the coast. Nothing came of it, except his book, *An Essay Explanatory of the Tempest Prognosticator* (London: 1851).

Leeches in a glass jar or bottle.

The leeches remain at the bottom during absolutely fine and calm wet weather. When a change in the former is approaching, they move steadily upwards many hours, even twenty-four, or rather more, in advance. If a storm is rapidly approaching, the leeches become very restless, rising quickly; while previous to a thunderstorm they are invariably much disturbed, and remain out of the water. When the change occurs and is passing over, they are quiet, and descend again. If under these circumstances they rise and continue above water, length or violence of storm is indicated. If they rise during

Leeches in a glass jar or bottle.

a continuance of east wind, strong winds rather than rain are to be looked for. When a storm comes direct from a distance, observe the rapid rising alluded to above, but much less notice is given, four to six hours. When heavy rains or strong winds are approaching, the leeches are restless, but their movements are less rapid, and they often remain half out of the water and quiet.—ELIZABETH WOOLLAMS.

A leech confined in a bottle of water is always agitated when a change of weather is about to take place. Before *high winds* it moves about with much celerity. Previous to *slight rain or snow* it creeps to the top of the bottle, but soon sinks; but if the *rain or wind* is likely to be of long duration, the leech remains a longer time at the surface. If *thunder* approaches, the leech starts about in an agitated and convulsive manner.

[SNAILS.] *Rain.*

When black snails cross your path,
Black cloud much moisture hath.

Snailie, snailie, shoot out your horn,
And tell us if it will be a bonnie day the morn.

If snails and slugs come out abundantly, it is a sign of rain.

When black snails on the road you see,
Then on the morrow rain will be.

Earth-worms.

If many earth-worms appear, it presages rain.

[GLOW-WORMS.] *Damp.*

When the glow-worm lights her lamp,
The air is always damp.

Rain.

Before rain
Glow-worms numerous, clear, and bright
Illume the dewy hills at night.—UNITED STATES.

If glow-worms shine much, it will rain.

Dry.

When the glow-worm glows, dry, hot weather follows.
UNITED STATES.

Reptiles, etc.

Serpent worship and rain.

Mr. Fergusson, in his *Tree and Serpent Worship*, states that "the chief characteristic of the serpents throughout the East in all ages seems to have been their power over the wind and rain, which they gave or withheld, according to their good or ill will towards man."

Snakes.

Hanging a dead snake on a tree will produce rain in a few hours.—NEGRO.

[*Note.*—Snakes are out before rain, and are therefore more easily killed.—MAJOR DUNWOODY.]

In Oregon the approach of snakes indicates that a spell of fine weather will follow. [SNAKES.]

When snakes are hunting food, rain may be expected; after a rain they cannot be found. *Hunting food.*

Snake-trails may be seen near houses before rain. *Trails.*

Rain is foretold by the appearance and activity of snakes. *Rain.*

When small water-snakes leave the sand in low, damp lands, frosts may be expected in three days.—APACHE INDIANS. *Water-snakes.*

Croaking frogs in spring will be three times frozen in. [FROGS.] *Croaking.*

When frogs warble, they herald rain.—ZUÑI INDIANS.

The louder the frog, the more the rain.

When frogs croak much, it is a sign of rain.

Yellow frogs are accounted a good sign in a hay-field, probably as indicating fine weather. *Yellow.*

If frogs make a noise in the time of cold rain, warm, dry weather will follow. *Noisy.*

If frogs, instead of yellow, appear russet green, it will presently rain. *Colour.*

Mr. Stroh informs me it was common to see in Germany and Switzerland a small green frog kept in a glass vessel half full of water, with a set of wooden steps leading down into the water; and the weather was supposed to be indicated by the position of the frog. If he remained in the water, fine weather was expected; if he emerged and sa upon the steps, rain and cold were indicated. *Green frog.*

When frogs spawn in the middle of the water, it is a sign of drought; and when at the side, it foretells a wet summer.
SCOTLAND. *Spawning.*

Tree-frogs piping during rain indicate a continuance. *Tree-frogs.*

Tree-frogs crawl up to the branches of trees before a change of weather. *Change.*

The green tree-frog becomes very unquiet before rain. *Rain.*

Tortoises creep deep into the ground, so as to completely conceal themselves from view, when a severe winter is to follow. *Tortoises.*

A salamander, kept in a bottle in the south of Spain, changed his position every day, and took up the most uncouth and extraordinary attitudes before a storm. *Salamander*

Almost any of the reptiles which pass the winter in a semi-dormant condition show signs by their attitude when any marked weather change ensues. *Reptiles.*

If toads come out of their holes in great numbers, rain will fall soon. *Toads.*

Insects.

Insects. The early appearance of insects indicates an early spring and good crops.—APACHE INDIANS.

[BEES.] *Returning.* When many bees enter the hive and none leave it, rain is near.—SCOTLAND.

Early. Bees early at work will not go on all day.

Swarm.
Bees will not swarm
Before a near storm.

Flight.
When bees to distance wing their flight,
Days are warm and skies are bright;
But when their flight ends near their home,
Stormy weather is sure to come.

Rain.
If bees stay at home,
Rain will soon come;

Fine.
If they fly away,
Fine will be the day.

At home.
When charged with stormy matter lower the skies,
The busy bee at home her labour plies;
Nor seeks the distant field and honeyed flower,
Returning laden'd with her golden store.
J. LAMB'S "ARATUS."

A bee was never caught in a shower.

[ANTS.] *Retiring.* Ants withdraw into their nests and busy themselves with their eggs before a storm.—THOMAS WILLSFORD.

Ants sometimes get down fifteen inches from the surface before very hot weather.—COMMUNICATED BY G. W. D. HANNAY.

Building.
If ants their walls do frequent build,
Rain will from the clouds be spilled.

Migration. When ants are situated in low ground, their migration may be taken as an indication of approaching heavy rains.

Travelling. Expect stormy weather when ants travel in lines, and fair weather when they scatter.

July. If in the beginning of July the ants are enlarging and building up their piles, an early and cold winter will follow.

Ant-hills open and closed. An open ant-hole indicates clear weather; a closed one, an approaching storm.

Active. If ants are more than ordinarily active, or if they remove their eggs from small hills, it will surely rain.

Wasps. Wasps building nests in exposed places indicate a dry season.

Wasps in great numbers and busy indicate warm weather.

Hornets. Hornets build nests high before warm summers, and low before cold and early winters.

Hornets.

When bounteous autumn crowns the circling year,
And fields and groves his russet livery wear,
If from the earth the numerous hornets rise,
Sweeping a living whirlwind through the skies,
Then close on autumn's steps will winter stern
With blustering winds and chilling rains return.
Pity the wretch who shelterless remains,
And the keen blast, half fed, half clad, sustains.
J. LAMB'S "ARATUS."

[SPIDERS.] *Busy.*

Spiders work hard and spin their webs a little before wind, as if desiring to anticipate it, for they cannot spin when the wind begins to blow.—BACON.

Rain.

Before rain or wind spiders fix their frame-lines unusually short. If they make them very long, the weather will usually be fine for fourteen days.

Indolent.

If the spiders are totally indolent, rain generally soon follows. Their activity during rain is a certain proof of its short duration. If they mend their webs between 6 and 7 p.m., it is a sign of a serene night.—J. W. G. GUTCH.

Changing webs.

Spiders generally change their webs once in every twenty-four hours. If they make the change between 6 and 7 p.m., expect a fair night. If they change their web in the morning, a fine day may be expected.

On walls.

Spiders, when they are seen crawling on the walls more than usually, indicate that rain will probably ensue. This prognostic seldom fails, particularly in winter.

Removing.

If spiders break off and remove their webs, the weather will be wet.

Renewing webs.

If spiders make new webs, and ants build new hills, the weather will be clear.

Working.

If the spider works during rain, it is an indication that the weather will soon be clear.

Cleaning.

When the spider cleans its web, fair weather is indicated.

Creeping out

Spiders creep out of their holes against wind and rain, Minerva having made them sensible of an approaching storm.
THOMAS WILLSFORD.

Mode of working.

If spiders in spinning their webs make the terminating filaments long, we may, in proportion to their lengths, expect rain.

Spiders' webs.

When spiders' webs in air do fly,
The spell will soon be very dry.

Spiders' webs. Dewy. Spiders' webs scattered thickly over a field covered with dew glistening in the morning sun indicate rain.

Long. Long, single, separate spiders' webs on grass indicate frost next night.—IRELAND.

Spiders' webs floating at autumn sunset
Bring a night frost—this you may bet.
UNITED STATES.

Garden spiders. If garden spiders forsake their cobwebs, rain is at hand.

If the garden spiders break and destroy their webs and creep away, expect continued rain.

Spiders in motion indicate rain.

Gossamer. When you see gossamer flying,
Be sure the air is drying.

Scorpions. When scorpions crawl, expect dry weather.

Tarantulas. When tarantulas crawl by day, rain will surely come.
CALIFORNIA.

Woodlice. If woodlice run about in great numbers, expect rain.

Harvest flies. When harvest flies hum,
Warm weather to come.

[HOUSE FLIES.] *Coming in house.* House flies coming into the house in great numbers indicate rain.

Rhyme. A fly on your nose, you slap, and it goes;
If it comes back again, it will bring a good rain.

Clinging. If flies cling much to the ceilings, or disappear, rain may be expected.

Seasons. If flies in the spring or summer grow busier or blinder than at other times, or are seen to shroud themselves in warm places, expect either hail, cold storms of rain, or much wet weather.

If in autumn the flies repair unto their winter quarters, it presages frosty mornings, cold storms, and the approach of winter. Atoms or small flies swarming together and sporting in the sunbeams give omen of fair weather.
THOMAS WILLSFORD.

Stinging. If flies sting and are more troublesome than usual, a change approaches.

Fall bugs. Fall bugs begin to chirp six weeks before a frost in the fall.
UNITED STATES.

Fleas. When fleas do very many grow
Then 'twill surely rain or snow.
When eager bites the thirsty flea,
Clouds and rain you sure shall see.

Butterflies The early appearance of butterflies is said to indicate fine weather.

When the white butterfly flies from the south-west, expect rain. *Butterflies.*

When the butterfly comes, comes also the summer.

ZUÑI INDIANS.

When the chrysalides are found suspended from the under side of rails, branches, etc., as if to protect them from rain, expect much rain. If they are found on slender branches, fair weather will last some time.—WESTERN PENNSYLVANIA. *Chrysalides.*

Fireflies in great numbers indicate fair weather. *Fireflies.*

If little flies or gnats be seen to hover together about the beams of the sun before it set, and fly together, making, as it were, the form of a pillar, it is a sure token of fair weather. [GNATS.] *In evening.*

HUSBANDMAN'S PRACTICE.

If gnats play up and down, it is a sign of heat; but if in the shade, it presages mild showers. If they collect in the evening before sunset, and form a vortex or column, fine weather will follow; while if they sting much, it is held to be an unfailing indication of rain. *Sporting.*

Gnats in October are a sign of long fair weather. *In October.*

Many gnats in spring indicate that the autumn will be warm. *Numerous.*

If gnats fly in large numbers, the weather will be fine.

If gnats bite sharper than usual, expect rain. *Biting.*

If gnats fly in compact bodies in the beams of the setting sun, expect fine weather. *Swarming.*

When locusts are heard, dry weather will follow, and frost will occur in six weeks.—UNITED STATES. *Locusts.*

When crickets chirp unusually, wet is expected. *Crickets.*

It is easy to foretell what sort of summer it will be by the position in which the larva of Cicada (*Aphrophora spumaria*) is found to lie in the froth (cuckoo spit) in which it is enveloped. If the insect lie with its head upwards, it infallibly denotes a dry summer; if downwards, a wet one. *Larva of Cicada.*

Before rain beetles and crickets are more troublesome than usual. *Beetles and crickets.*

The clock beetle, which flies about in the summer evenings in a circular direction, with a loud, buzzing noise, is said to foretell a fine day. It was consecrated by the Egyptians to the sun.—C. SWAINSON. *Clock beetle.*

If the clock beetle flies circularly and buzzes, it is a sign of fine weather.

A certain long-bodied beetle is called in Bedfordshire the "rain beetle," on account of its always appearing before rain. *Rain beetle.*

When little black insects appear on the snow, expect a thaw. *Black insects*

Plants, etc.

The vegetable world has not escaped the notice of the weather prophets, and many plants have been observed to give indications of stormy weather long before it actually takes place. The closing, for instance, of the pink-eyed pimpernel, or ploughman's weather glass, is better understood among the Bedfordshire labourers than the indications of any instrument, and has to them the great advantage of being in the fields where they work, of being easily understood, and of costing nothing. From the blossoming and fruition of certain plants a rough code of rules has also been laid down as to the coming harvest, the time for sowing, and the severity or mildness of the seasons. These will be found mentioned in their proper places.

Trees. Trees snapping and cracking in the autumn indicate dry weather.

[LEAVES.] *Rattling.* When dry leaves rattle on the trees, expect snow.

Turning.

When the leaves show their under sides,
Be very sure that rain betides.

Curling. When the leaves of trees curl with the wind from the south, it indicates rain.

Falling. The leaves of trees fall sooner on the south side; but vine shoots burst out on that side, and have scarce any other aspect.—PLINY.

Remaining.

If on the trees the leaves still hold,
The coming winter will be cold.

Flying. Leaves and straws playing in the air when no breeze is felt, the down of plants flying about, and feathers floating and playing on the water, show that winds are at hand.—BACON.

[FLOWERS.] The odour of flowers is more apparent just before a shower (when the air is moist) than at any other time.

Early. Early blossoms indicate a bad fruit year.

Date of plants flowering. Miss Ormerod, F.R.Met.Soc., has noticed that bulbous and surface-rooted plants have wider differences as to the date of first flowering than the deeper-rooted plants. This is on account of the deep-rooted plants being slower to acquire the temperature of the air.

Dead branches. Dead branches falling in calm weather indicate rain.

Berries. Plenty of berries indicates a severe winter.

[OAK AND ASH.] When the oak comes out before the ash, there will be fine weather in harvest; but when the ash comes out before the oak, the harvest will be wet.—MIDLAND COUNTIES.

[*Oak and ash.*]
Budding.

When the ash is out before the oak,
Then we may expect a choke [drought];
When the oak is out before the ash,
Then we may expect a splash [rain].
SHROPSHIRE.

If the oak's before the ash,
Then you'll only get a splash;
But if the ash precedes the oak,
Then you may expect a soak.

When buds the oak before the ash,
You'll only have a summer splash.

The ash before the oak,
Choke, choke, choke;
The oak before the ash,
Splash, splash, splash.
[Contradicting the former.]

If buds the ash before the oak,
You'll surely have a summer soak;
But if behind the oak the ash is,
You'll only have a few light splashes.

Mr. Douglas, of Babworth, says that the oak is always in leaf before the ash, if the subsoil is in a moist state.

[A correspondent to *Notes and Queries*, June 21st, 1873, says he never knew the ash to come into leaf before the oak.]

If the ash is out before the oak,
You may expect a thorough soak;
If the oak is out before the ash,
You'll hardly get a single splash.

Oak, smoke [summer hot].
Ash, squash [summer wet].

If the oak is out before the ash,
'Twill be a summer of wet and splash;
But if the ash is before the oak,
'Twill be a summer of fire and smoke.
HAMPSHIRE.

[OAK. *Gall.*

The oak gall is examined by the Spanish peasants when the wheat is in ear. If they find a maggot, they say the harvest will be good; if an insect already hatched, the contrary.

Budding.

You must look for grass on the top of an oak tree.

[*i.e.*, the grass seldom springs well till the oak comes out.

Fruitful.

If the oak bear much mast [acorns], it foreshows a long and hard winter.—WORLEDGE.

[Oak.]

When the oak puts on his gosling grey,
'Tis time to sow barley, night or day.

Oak apples. There is a superstition about examining the oak apples on September 29th, and auguries are inferred from their condition. See *Husbandman's Practice; or, Prognostication For Ever.*

QUOTED BY C. SWAINSON IN "WEATHER FOLK-LORE."

Ash. Black as ash buds in the front of March.—TENNYSON.

Beech and oak. When beech mast thrives well, and oak trees hang full, a hard winter will follow, with much snow.

Beech nuts. When beech nuts are plentiful, expect a mild winter.

Elm leaves and barley sowing.

When the elmen leaf is as big as a mouse's ear,
Then to sow barley never fear.
When the elmen leaf is as big as an ox's eye,
Then says I, "Hie, boys! hie!"

"FIELD," APRIL 28TH, 1866.

Elm leaves and kidney beans.

When elm leaves are as big as a shilling,
Plant kidney beans, if to plant 'em you're willing;
When elm leaves are as big as a penny,
You must plant kidney beans, if you mean to have any.

WORCESTERSHIRE.

Silver maple. The silver maple shows the lining of its leaf before a storm.

UNITED STATES.

Sugar maple. When the leaves of the sugar maple tree are turned upside-down, expect rain.—UNITED STATES.

Pine. Pine cones hung up in the house will close themselves against wet and cold weather, and open against hot and dry times.

THOMAS WILLSFORD.

Mulberry. When the mulberry has shown green leaf, there will be no more frost.—GLOUCESTERSHIRE.

When the mulberry buds and puts forth its leaves, fear no frosts or bad weather.—PLINY.

Almond tree.

Mark well the flowering almonds in the wood:
If odorous blooms the bearing branches load,
The glebe will answer to the sylvan reign,
Great heats will follow, and large crops of grain;
But if a wood of leaves o'ershades the tree,
Such and so barren will the harvest be.

VIRGIL'S "GEORGICS."

Sloe tree.

When the sloe tree is white as a sheet,
Sow your barley, whether it be dry or wet.

Cottonwood and quaking asp. Cottonwood and quaking asp trees turn up their leaves before rain.—UNITED STATES.

Trembling of aspen leaves in calm weather indicates an approaching storm.—UNITED STATES. *Aspen leaves.*

Before rain the leaves of the lime, sycamore, plane, and poplar trees show a great deal more of their under surfaces when trembling in the wind. *Leaves.*

[*Note.*—This is because the damp air softens the leat stalks.]

When the blooms of the dogwood tree are full, expect a cold winter; when the blooms of the same are light, expect a warm winter. *Dogwood blossoms.*

Frost will not occur after the dogwood blossoms.

UNITED STATES.

You may shear your sheep
When the elder blossoms peep. *Elder blossom.*

Witches were thought to produce bad weather by stirring water with branches of elder. *Elder bush.*

When cockle burs mature brown, it indicates frost. *Cockle burs.*

UNITED STATES.

Its always cold when the hawthorn blossoms. *Hawthorn.*

Harvest follows in thirteen weeks after the milk-white thorn scents the air.—SCOTCH. *Thorn.*

If many whitethorn blossoms or dog-roses are seen, expect a severe winter. *Whitethorns and dog-roses.*

When the bramble blossoms early in June, an early harvest is expected.—SCOTLAND. *Bramble.*

Dead nettles in abundance late in the year are a sign of a mild winter.—UNITED STATES. *Dead nettle.*

Just before rain or heavy dew the wild indigo closes or folds its leaves.—UNITED STATES. *Wild indigo.*

Corn (Indian) fodder dry and crisp indicates fair weather; but damp and limp, rain. It is very sensitive to hygrometric changes. *Corn fodder*

Ears of corn (Indian) are said to be covered with thicker and stronger husks before hard winters. *Corn husks.*

If corn (maize) is hard to husk, expect a hard winter.

APACHE INDIANS.

Make hay while the sun shines. *Hay.*

A double husk on corn (maize) indicates a severe winter.

Corn in good years is hay; in ill years straw is corn. *Corn and hay.*

T. FULLER.

Sow wheat in dirt, and rye in dust. *Wheat and rye.*

Wild oat. A beard of wild oat, with its adhering capsule, fixed on a stand, serves the purpose of a hygrometer, twisting itself more or less, according to the moisture of the air.—E. DARWIN.

Hay and buckwheat. If the hay is black (with wet), the buckwheat will be white (with blossom).—RUSSIA.

[BEANS.] *Sowing.* Plant garden beans when the sign is in the scales; they will hang full.

Sow beans in the mud,
And they'll grow like a wood.

Plant the bean when the moon is light;
Plant potatoes when the moon is dark.

Whitlow grass. We may look for wet weather if the leaves of the whitlow grass (*Draba verna*) droop, and if lady's bedstraw (*Galium verum*) becomes inflated and gives out a strong odour.

Sensitive plants. Sensitive plants contract their leaves at the approach of rain.

Abrus precatorius. The so-called "weather plant" is said by some to foretell the weather for an enormous area by the behaviour of its leaves, which when horizontal indicate change: if they slope upwards, fine weather; but if they droop, bad weather is to be expected.

Daffodils.

Daffodils
That come before the swallow dares, and take
The winds of March with beauty.—SHAKESPEARE.

[DANDELIONS.] When the down of the dandelion contracts, it is a sign of rain.

Down. If the down flyeth off colt's-foot, dandelyon, and thistles, when there is no winde, it is a signe of rain.—COLES.

Closing. The dandelions close their blossoms before a storm; the sensitive plant its leaves. The leaves of the may tree bear up, so that the under side may be seen before a storm.

Blooming. When the dandelions bloom early in spring, there will be a short season. When they bloom late, expect a dry summer.

Wood sorrel. A species of wood sorrel contracts its leaves at the approach of rain.

Trefoil. The stalk of trefoil swells before rain.—BACON.

Pliny mentions it as a fact that trefoil bristles and erects its leaves against a storm.—BACON.

Wood anemone and swallows. In Sweden the wood anemone begins to blow on the arrival of the swallow.—LINNÆUS.

The yellow wood anemone and the wind flower (*Anemone nemorosa*) close their petals and droop before rain.

Wood anemone. The wood anemone never opens its petals but when the wind blows, whence its name.

[CLOVER.] Clover contracts its leaves at the approach of a storm.

Rough. When clover grass looks rough, and its leaves stand staring up, it is a sign of a tempest.—PLINY.

Clover grass is rough to the touch when stormy weather is at hand.

[ONIONS.] *Skin.* When the onion's skin is thin and delicate, expect a mild winter; but when the bulb is covered by a thick coat, it is held to foreshow a severe season.

> Onion's skin very thin,
> Mild winter coming in;
> Onion's skin thick and tough,
> Coming winter cold and rough.
> GARDENER'S RHYME.

Hedge fruit.

> Mony haws,
> Mony snaws;
> Mony slaes,
> Mony cold taes.—SCOTLAND.

> Mony hips and haws,
> Mony frosts and snaws.—SCOTLAND.

Broom. The broom having plenty of blossoms is a sign of a fruitful year of corn.—THOMAS WILLSFORD.

Fern. It was anciently supposed that the burning of fern drew down the rain.

Mountain ash.

> Mony rains, mony rowans; *
> Mony rowans, mony yawns.†—SCOTLAND.

[CHICKWEED.] *Expanding.* Chickweed expands it leaves boldly and fully when fine weather is to follow; but if it should shut up, then the traveller is to put on his great coat.

Half opening. The half opening of the flowers of the chickweed is a sign that the wet will not last long.

Night. If the flowers keep open all night, the weather will be wet next day.

Siberian sow-thistle. The non-closing of the flower-heads of the sow-thistle warns us that it will rain next day, whilst the closing of them denotes fine weather.

Convolvulus. The convolvulus folds up its petals at the approach of rain.

* Rowans are the fruit of the mountain ash.

† Yawns are light grains of wheat, oats, or barley.

African marigold. If this plant do not open its petals by seven in the morning, it will rain or thunder that day. It also closes before a storm.

Cape marigold. If the small Cape marigold (*Calendula pluvialis*) should open at six or seven in the morning, and not close till four in the afternoon, we may reckon on settled weather.

Marsh marigold and cuckoo. The marsh marigold blows when the cuckoo sings.
STILLINGFLEET IN ENGLAND; LINNÆUS IN SWEDEN.

Marigold.

The marigold that goes to bed with the sun,
And with him rises, weeping.—SHAKESPEARE.

Seaweed. A piece of kelp or seaweed hung up will become damp previous to rain.

Pink-eyea pimpernel. When this flower closes in the daytime, it is a sign of rain.*

Pimpernel, pimpernel, tell me true
Whether the weather be fine or no;
No heart can think, no tongue can tell
The virtues of the pimpernel.
"FOLK-LORE JOURNAL," 1889.

Now, look! Our weather glass is spread—
The pimpernel, whose flower
Closes its leaves of spotted red
Against a rainy hour.—PROFESSOR WILSON.

Teasel. Teasel or Fuller's thistle hung up will open for fine weather, and close for wet.—THOMAS WILLSFORD.

Flowers. The bladder-ketmir, the stemless ground thistle, the marsh marigold, the creeping crowfoot, the wood sorrel, foreshow the weather in various ways—viz., when the flowers of the first do not open, when the second closes its calyx, and when the rest fold their leaves.—MR. HANNEMAN, OF PROSKAU.

Cowslip. The cowslip stalks being short are said to foreshow a dry summer.

Gentian. The gentian (*Gentiana pneumonanthe*) closes up both flowers and leaves before rain.

Burnet. The burnet saxifrage (*Pimpinella saxifraga*) indicates by half opening its flowers that the rain is soon to cease.

Toadstools. If toadstools spring up in the night in dry weather, they indicate rain.

Thistles.

Cut 'em in June, they'll come again soon;
Cut 'em in July, they *may* die;
Cut 'em in August, die they must.—SHROPSHIRE.

* This flower is known as the ploughman's weather glass.

Not signless by the husbandmen are seen
The Ilex and Lentiscus darkly green.
If an abundant crop the Ilex bear,
With blighting matter teems the vapoury air;
If with unusual weight its branches groan,
Then their light sheaves the hapless farmers moan.
J. LAMB'S "ARATUS."

Ilex and Lentiscus.

Thrice in the course of each revolving year
On the Lentiscus flowers and fruit appear;
And three convenient times to farmers show
To break the fertile clod with crookèd plough.
If at each time this tree with fruit abound,
Each time with stores will teem the fruitful ground.
And like prognostic yields the humble squill,
Thrice flowering yearly by the purple rill.
J. LAMB'S "ARATUS."

Lentiscus.

Squill.

The indications of plants as to the times for sheep shearing, harvest, etc., will be found under the head of "Times and Seasons."

Various plants.

Various.

Bacon tried an experiment, and found that four ounces of wool let down a well, yet not so as to touch the water, increased to five ounces and one dram in weight during one night (by the moisture).

[WOOL.] *Damp experiment.*

[*Note.*—Vitruvius, the architect, mentions a similar experiment made in a small pit, in order to see whether it was a good place to sink further for water.]

A fleece of wool by lying long on the ground gains weight, which could not be unless something pneumatic were condensed into something ponderable. In ancient times sailors used to cover the sides of ships at night with fleeces of wool like coverlets or curtains, but not so as to touch the water; and in the morning they would squeeze out of them fresh water for use on the voyage.—BACON.

Collecting fresh water at sea.

When rheumatic people complain of more than ordinary pains in the joints, it will rain.

Rheumatism.

As old sinners have all points
O' th' compass in their bones and joints—
Can by their pangs and aches find
All turns and changes of the wind,
And better than by Napier's bones *
Feel in their own the age of moons.—BUTLER.

* Certain engraved slips invented by Napier to facilitate calculations.

Muscæ volitantes. The deceptive appearance of motes or small flies moving before the eyes is said to presage rain and storms.

Stomach. In persons of weak and irritable constitution the digestive powers are much influenced by the weather. Before storms such persons are uneasy.

Scalp locks. When the locks of the Navajoes turn damp in the scalp-house, surely it will rain.

Dreams. Dreams of a hurrying and frightful nature, and imperfect sleep, are frequent indications that the weather has changed, or is about to change. Many persons experience these nocturnal symptoms on a change of wind, particularly when it becomes easterly.

Corns, wounds, and sores. If corns, wounds, and sores itch or ache more than usual, rain is to fall shortly.

Corns.

A coming storm your shooting corns presage,
And aches will throb, your hollow tooth will rage.

BROOME.

Ears. Ringing in the ear at night indicates a change of wind.

A singing in the ears sometimes indicates a change of weather, generally an increase of pressure or rise in the barometer.

Appetite. When everything at the table is eaten, it indicates continued clear weather.

Cream and milk. Cream and milk, when they turn sour in the night, often indicate thereby that thunderstorms are about.

Superstitions. The presence of a dead body on ship or boat is supposed to cause contrary winds. Eggs are credited with the same power. So is whistling.—FOLK-LORE JOURNAL.

Chairs and tables. When chairs and tables creak and crack, it will rain.

Doors, etc. Doors and windows are hard to shut in damp weather.

Floors. Oiled floors become very damp before rain.

Wooden hygroscope. Mr. Edgworth is mentioned as having made a wooden automaton, consisting of a long slip of wood cut crosswise to the grain, and furnished with two points at each end pointing backwards, thus—

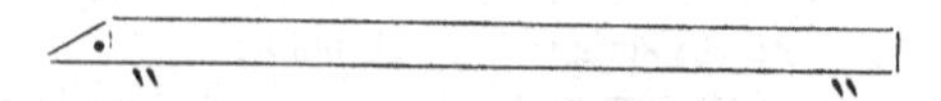

The effect of this was that when the wooden figure expanded with the dampness of the air, it pushed forward its head; and when it shrank in drying, it dragged its tail; so that it continually went forward according to the dampness of the season, and the distance passed gave a rough indication of the comparative moisture of the air.—E. DARWIN.

Camphor gum dissolved in alcohol is said to throw out feathery crystals and to rise before rain. *Camphor gum.*

If the matting on the floor is shrinking, dry weather may be expected. When the matting expands, expect wet weather. *Matting on floor.*

The sailor notes the tightening of the cordage of his ship as a sign of coming rain. *Cordage.*

Stringed instruments giving forth clear, ringing sounds indicate fair weather. *Stringed instruments.*

Strings ot catgut or whipcord untwist and become longer during a dry state of the air, and *vice versâ.* *Strings, etc.*

On this principle is constructed the weather-house—a toy usually found in country houses, and from which the figure of a woman emerges in fine weather, while a man wrapped in a great coat comes out before rain. *Weather house-toy.*

A lump of hemp acts as a good hygrometer. *Hemp.*

Ropes being difficult to untwist indicate bad weather. *Ropes.*

Before wind and rain, it is said that the black damp extinguishing the lights is observed at the bottom of ironstone pits and through the "waste."—SIR A. MITCHELL. *Mines damp.*

In Midlothian the miners think that approaching changes of the weather are preceded by an increased flow of water and the issue of gases and foul air from the crevices; and when very bad weather is at hand, these last escape with a characteristic sound like the buzz of insects.—SIR A. MITCHELL. *Flow of water in mines.*

Quarries of stone and slate indicate rain by a moist exudation from the stones. *Quarries.*

A stone in Finland, called the "weather stone" (doubtless saturated with salt water), breaks out into dark spots at the approach of rain. Mr. A. Whittaker says the stone is a fossil containing clay, rock-salt, and nitre.—ENGLISH MECHANIC. *Damp stones.*

When walls built of stones which have been quarried below high-water mark become damp, wet weather is at hand. *Stones.*

If any one sits on a stone (one of the Stiper Stones) called the Devil's Chair, a thunderstorm immediately arises.

SHROPSHIRE SUPERSTITION.

When walls are more than usually damp, rain is expected. *Walls.*

If stoves or iron rust during the night, it is a sign of rain. *Iron.*

Salt increases in weight before a shower. *Salt.*

A farmer's wife says, when her cheese salt is soft, it will rain; when getting dry, fair weather may be expected.

Salt. There is a pillar of salt in the mines of Cracow which is called "Lot's Wife," and which becomes damp at the approach of rain.

Earthquakes. Previous to earthquakes, the orb of the sun is of an unusua colour—remarkably red, or tending to black. Bodies are seen running in the heavens, accompanied with abundance of flame, and the stars appear of a shape different from that which they possessed before.

"PAUSANIAS," TAYLOR'S TRANSLATION, 1794.

Soap. Soap covered with moisture indicates bad weather.

Dust. Dust rising in dry weather is a sign of approaching change.

SCOTLAND.

Eddies. If dust whirls round in eddies when being blown about by the wind, it is a sign of rain.

Dust at sea. A curious phenomenon has frequently been observed to accompany northerly winds, which is: that in March or April ships that are bound to Bombay or Surat frequently have their rigging covered with white dust, although several degrees distant from Canara or Concan. The northerly and north-north-west winds, blowing from the coast of Persia, over an extensive surface of sea (at least ten or twelve degrees), it is difficult to judge what can occasion the dust, if it is not generated in the atmosphere, which is in these months sometimes impregnated with a dry haze.

J. HORSBURGH IN "NICHOLSON'S JOURNAL."

Kites. If kites fly high, fine weather is at hand.

Smoke. Smoke falling to the ground indicates rain.

When the smoke of the Tharsis mine (Spain) blows northward, it is a sign of rain.

If, during calm, smoke does not ascend readily, expect rain.

Tobacco smoke. If the smoke of a morning pipe hangs a long while in the air a good hunting day always follows.

Tobacco pipes. When the odour of pipes is longer retained than usual, and seems denser and more powerful, it often forebodes rain and wind.

Spectroscope. Moisture in the air is shown by a dark line in the spectrum near the D lines, which latter are almost fused together when the "rain-band" is very marked.

Bladder. Pig's bladder, when stretched, fine; when flaccid, wet.

Pavements. If pavements appear rusty, rain will follow.

Pliny asserts that vessels containing eatables sometimes leave a sweat behind them in the storerooms, and that this is a sign of fearful storms.—BACON. *Earthen vessels.*

If metal plates and dishes sweat, it is a sign of bad weather. *Plates.*
PLINY.

When the sparks stick to the poker, it is a sign of rain. *Sparks.*
SPANISH.

When the flames of candles flare and snap or burn with an unsteady or dim light, rain and frequently wind also are found to follow. *Candles.*

Excrescences forming about wicks of lamps and candles, which consume their fuel slowly, indicate rain. *Lamp wicks.*

[COALS.]

Coals, when they burn very bright, foretell wind, and likewise when they quickly cast off and deposit their ashes.—BACON. *Bright.*

If the burning coals stick to the bottom of the pot, it is a sign of a tempest.—PLINY. *Adhering.*

Coals covered with thick white ashes indicate snow in winter, and rain in summer. *Ashy.*

Coals becoming alternately bright and dim indicate approaching storms. *Flaming.*

[FIRES.]

Burning wood in winter pops more before snow.

Fires burning paler than usual and murmuring within are significant of storms. If the flame shoot in a twisting and curling form, it principally denotes wind; but fungous growths or excrescences on the wicks of lamps rather foreshadow rain. *Flames.*
BACON.

Fire is said to burn brighter and throw out more heat just before a storm. *Bright.*

If the fire burns unusually fiercely and brightly in winter, there will be frost and clear weather; if the fire burns dull, expect damp and rain. *Fierce.*

Blacksmiths select a stormy day in which to perform work requiring extra heat. *In storm.*

A fire hard to kindle indicates bad weather. *Difficult to light.*

When the fire crackles lightly, it is said to be treading snow. *Crackling.*
OLD WOMAN.

An empty Florence oil flask inverted, with the open neck placed in a glass of water, is sometimes used as a barometer, the level of the water in the neck being high for good weather, and *vice versâ*. *Bottle.*

Ashes.

But why abroad to seek prognostics go,
When ashes vile foretell the falling snow,
When half consumed the coals to cinders turn,
And with a sputtering flame the torches burn?
And hail expect when the burnt cinders white
With glowing heat send round a glaring light.

J. LAMB'S "ARATUS."

Torches, etc.

When the dull fire emits no cheerful rays,
With lustre dimmed the languid torches blaze,
And the light cobwebs float along the air—
No symptoms these of weather calm and fair.

J. LAMB'S "ARATUS."

[SOOT.]

If soot falls down the chimney, rain will ensue.

Burning.

Soot burning on the back of the chimney indicates storms. When the soot sparkles on pots over the fire, rain follows.

Stagnant water.

If standing water be at any time warmer than it was commonly wont to be, and no sunshine help, it foretelleth rain.

HUSBANDMAN'S PRACTICE.

Springs.

Springs running flusher (commonly called "earth sweat") is an indication of rain.

Creeks and springs.

In dry weather, when creeks and springs that have gone dry become moist, or, as we may say, begin to sweat, it indicates approaching rain. Many springs that have gone dry will give a good flow of water just before rain.

J. E. WALTER (KANSAS).

Springs rising.

Springs rise against rain.

Ditches, drains, etc.

Drains, ditches, and dunghills are more offensive before rain.

Coffee bubbles.

When the bubbles of coffee collect in the centre of the cup, expect fair weather. When they adhere to the cup, forming a ring, expect rain. If they separate without assuming any fixed position, expect changeable weather.

Various signs of bad weather.

The changing weather certain signs reveal;
Ere winter sheds her snow or frosts congeal,
You'll see the coals in brighter flame aspire,
And sulphur tinge with blue the rising fire.
You'll hear the sounds
Of whistling winds ere kennels break their bounds;
Ungrateful odours common shores diffuse,
And dropping vaults distil unwholesome dews,
Ere the tiles rattle with the smoking shower.

Let credulous boys and prattling nurses tell
How, if the festival of Paul be clear,
Plenty with liberal horn shall strew the year;
When the dark skies dissolve in snow or rain,
The labouring hind shall yoke the steer in vain;
But if the threatening winds in tempests roar,
The war shall bathe her wasteful sword in gore.

Various signs of bad weather.

GAY.

No weather fair expect, when Iris throws
Around the azure vault two painted bows;
When a bright star in night's blue vault is found
Like a small sun by circling halo bound;
When dip the swallows as the pool they skim,
And waterfowls their ruffled plumage trim;
When loudly croak the tenants of the lake,
Unhappy victims of the hydra snake;
When at the early dawn from murmuring throat
Lone Ololygo pours her dismal note;
When the hoarse raven seeks the shallow waves—
Dips her black head—her wings and body laves.
The ox looks up and snuffs the coming showers
Ere yet with pregnant clouds the welkin lowers;
Dragging from vaulted cave their eggs to view,
Th' industrious ants their ceaseless toil pursue;
While numerous insects creep along the wall,
And through the grass the slimy earth-worms crawl—
The black earth's entrails men these reptiles call.
Cackles the hen as sounds the dripping rill,
Combing her plumage with her crookèd bill.

Double rainbow. Halo. Swallows. Waterfowls. Frogs. Owl? Raven. Ox. Ants. Insects. Worms. Hen.

* * * * *

When flocks of rooks or daws in clouds arise,
Deafening the welkin with discordant cries;
When from their throats a gurgling note they strain,
And imitate big drops of falling rain;
When the tame duck her outstretched pinion shakes;
When the shrill, screaming hern the ocean seeks,—
All these prognostics to the wise declare
Pregnant with rain, though now serene, the air.

Rooks, daws. Duck. Hern.

* * * * *

No weather calm expect, when, floating high,
Cloud rides o'er cloud; when clamorous cry
The geese; when through the night the raven caws,
And chatter loud at eventide the daws;
When sparrows ceaseless chirp at dawn of day,
And in their holes the wren and robin stay.

Clouds. Geese, raven. Daws. Sparrows. Wren, robin.

* * * * *

Wild ducks. When from their briny couch the wild ducks soar,
And beat with clanging wings the echoing shore;
Clouds on mountains. When gathering clouds are rolled as drifting snow
In giant length along the mountain's brow;
Thistle-down, or foam of that appearance. When the light down that crowns the thistle's head
On ocean's calm and glassy face is spread,
Extending far and wide,—the sailors hail
These signs prophetic of the rising gale.

* * * * *

Waves. When the long, hollow, rolling billows roar,
Breaking in froth upon the echoing shore;
And through the rugged rock and craggy steep
Sound of waves. Whispers a murmuring sound, not loud, but deep;
Hern. When screaming to the land the lone hern flies,
And from the crag reiterates her cries;
Sea-mews. Breasting the wind in flocks the sea-mews sail,
And smooth their plumes against th' opposing gale;
Cormorants. And diving cormorants their wings expand,
And tread—strange visitors—the solid land.
[Signs of bad weather.]

* * * * *

Crab. Before the storm the crab his briny home
Sidelong forsakes, and strives on land to roam;
Mice. The busy household mice shake up with care
Their strawy beds, and for long sleep prepare.
Flies. When keen the flies, a plague to man and beast,
Seek with proboscis sharp their bloody feast;
When in the wearisome, dark, wintry night
Torches. The flickering torches burn with sputtering light,
Now flaring far and wide, now sinking low,
While round their wicks the fungous tumours grow;
Embers. When on the hearth the burning ember glows,
And numerous sparks around the charcoal throws,—
Mark well these signs, though trifling, not in vain,
Prognostics sure of the impending rain.

* * * * *

Lamps. When burn the lamps with soft and steady light,
Owl. And the owl softly murmurs through the night;
Raven. And e'en the raven from her varying throat
Utters at eve a soft and joyous note;
When from all quarters in the twilight shade
Rooks. The rooks, returning to th' accustomed glade,
Their lofty rocking dormitories crowd,
Clapping their gladsome wings and cawing loud.
[Fine weather signs]

J. LAMB'S "ARATUS."

Animal prognostics of the weather.

When the small birds prune the wing,
Ducking in the limpid spring,
Languid 'neath the sheltering trees
Oxen snuff the southern breeze,
Cackling geese with outstretched throat
Join the crow's discordant note,
Busy moles throw up the earth,
Crickets chirrup on the hearth,
Loudly caws the harsh-toned rook,
Spotted frogs respondent croak,
Gnats wheel round in airy ring,
Angry wasps and hornets sting,
Cautious bees forbear to roam,
Honey seeking near their home,
Spiders from their cobwebs fall,
Forth the shiny earth-worms crawl,
Loud, sonorous asses bray,
Frequent crows the bird of day,
Hens and chicks run helter-skelter—
These, though cloudless be the sky,
Tokens are that rain is nigh.

Various signs of bad weather.

To him the wary Pilot thus replies:
A thousand omens threaten from the skies;
A thousand boding signs my soul affright,
And warn me not to tempt the seas this night.
In clouds the setting sun obscured his head,
Nor painted o'er the ruddy west with red:
Now north, now south, he shot his parted beams,
And tipped the sullen black with golden gleams.
Pale shone his middle orb with faintish rays,
And suffered mortal eyes at ease to gaze.
Nor rose the silver queen of night serene;
Supine and dull her blunted horns were seen,
With foggy stains and cloudy blots between.
Dreadful awhile she shone all fiery red,
Then sickened into pale, and hung her drooping head.
Nor less I fear from that hoarse, hollow roar
In leafy groves and on the sounding shore.
In various turns the doubtful dolphins play,
And thwart, and run across, and mix their way.
The cormorants the watery deeps forsake,
And soaring herns avoid the plashy lake;
While waddling on the margin of the main,
The crow bewets her, and prevents * the rain.

[The Pilot is addressing Cæsar, who wants to cross the Adriatic Gulf to Brundusium.]

LUCAN'S "PHARSALIA," V., ROWE'S TRANSLATION.

* = Goes before (old English).

Winds. The hollow winds begin to blow,
Clouds, barometer. The clouds look black, the glass is low,
Soot. The soot falls down, the spaniels sleep,
Spiders. And spiders from their cobwebs creep.
Sunset. Last night the sun went pale to bed,
Moon. The moon in haloes hid her head,
The boding shepherd heaves a sigh,
Rainbow. For, see! a rainbow spans the sky;
Walls, ditches. The walls are damp, the ditches smell,
Pimpernel. Closed is the pink-eyed pimpernel;
Chairs. Hark how the chairs and tables crack!
Joints. Old Betty's joints are on the rack;
Ducks, peacocks, hills. Loud quack the ducks, the peacocks cry,
The distant hills are looking nigh;
Swine. How restless are the snorting swine!
Flies. The busy flies disturb the kine;
Swallow. Low o'er the grass the swallow wings;
Cricket. The cricket, too, how sharp he sings!
Cat. Puss on the hearth, with velvet paws,
Sits wiping o'er her whiskered jaws;
Fishes. Through the clear stream the fishes rise,
And nimbly catch the incautious flies;
Glow-worms. The glow-worms, numerous and bright,
Illumed the dewy dell last night;
Toad. At dusk the squalid toad was seen
Hopping and crawling o'er the green;
Dust. The whirling dust the wind obeys,
And in the rapid eddy plays;
Frog. The frog has changed his yellow vest,
And in a russet coat is dressed;
Air. Though June, the air is cold and still,
Blackbird. The merry blackbird's voice is shrill;
Dog. My dog, so altered in his taste,
Quits mutton bones on grass to feast;
Rooks. And see yon rooks, how odd their flight!
They imitate the gliding kite,
And seem precipitate to fall,
As if they felt the piercing ball.
'Twill surely rain,—I see with sorrow
Our jaunt must be put off to-morrow.

DR. E. DARWIN; ALSO ATTRIBUTED TO DR. JENNER.

[WIND.] For ere the rising winds begin to roar,
Sea. The working seas advance to wash the shore,
Leaves. Soft whispers run along the leafy woods,

And mountains whistle to the murmuring floods. *Mountains.*
Even then the doubtful billows scarce abstain *Waves.*
From the tossed vessel on the troubled main;
When crying cormorants forsake the sea, *Cormorants.*
And, stretching to the covert, wing their way;
When sportful coots run skimming o'er the strand; *Coots.*
When watchful herons leave their watery stand, *Herons.*
And, mounting upward with erected flight,
Gain on the skies, and soar above the sight:
And oft, before tempestuous winds arise,
The seeming stars fall headlong from the skies, *Meteors.*
And, shooting through the darkness, gild the night
With sweeping glories and long trails of light;
And chaff with eddy winds is whirled around, *Chaff.*
And dancing leaves are lifted from the ground; *Leaves.*
And floating feathers on the waters play: *Feathers.*
But when the wingèd thunder takes his way *Thunder.*
From the cold north, and east and west engage, *Winds.*
And at their frontiers meet with equal rage,
The clouds are crushed; a glut of gathered rain *Clouds.*
The hollow ditches fills, and floats the plain;
And sailors furl their dropping sheets amain.

Wet weather seldom hurts the most unwise; [RAIN.]
So plain the signs, such prophets are the skies.
The wary crane foresees it first, and sails *Crane.*
Above the storm, and leaves the lowly vales;
The cow looks up, and from afar can find *Cow.*
The change of heaven, and snuffs it in the wind;
The swallow skims the river's watery face; *Swallow.*
The frogs renew the croaks of their loquacious race; *Frogs.*
The careful ant her secret cell forsakes, *Ants.*
And drags her eggs along the narrow tracks;
At either bourn the rainbow drinks the flood; *Rainbow.*
Huge flocks of rising rooks forsake their food, *Rooks.*
And, crying, seek the shelter of the wood.
Besides the several sorts of watery fowls *Waterfowl.*
That swim the seas or haunt the standing pools,
The swans that sail along the silvery flood, *Swans.*
And dive with stretching necks to search their food,
Then lave their backs with sprinkling dews in vain,
And stem the stream to meet the promised rain,
The crow with clam'rous cries the shower demands, *Crow.*
And single stalks along the desert sands.
The nightly virgin, while her wheel she plies,
Foresees the storm impending in the skies,
When sparkling lamps their splutt'ring light advance, *Lamps.*
And in the sockets oily bubbles dance.

[FINE WEATHER.]	Then after showers 'tis easy to descry Returning suns and a serener sky.
Stars.	The stars shine smarter; and the moon adorns,
Moon.	As with unborrowed beams, her sharpened horns;
Gossamer.	The filmy gossamer now flits no more,
Kingfishers.	Nor halcyons bask on the short sunny shore;
Swine.	Their litter is not tossed by sows unclean;
Mist.	But a blue droughty mist descends upon the plain;
Owls.	And owls that mark the setting sun declare A starlight evening and a morning fair.
Hawk and lark.	Tow'ring aloft, avenging Nisus flies, While dared below the guilty Scylla lies. Wherever frighted Scylla flies away, Swift Nisus follows and pursues his prey; Where injured Nisus takes his airy course, Thence trembling Scylla flies and shuns his force. This punishment pursues the unhappy maid, And thus the purple hair is dearly paid.
Ravens.	Then thrice the ravens rend the liquid air, And croaking notes proclaim the settled fair. Then round their airy palaces they fly To greet the sun; and seized with secret joy, When storms are overblown, with food repair To their forsaken nests and callow care. Not that I think their breasts with heavenly souls Inspired, as man who destiny controls; But with the changeful temper of the skies, As rains condense and sunshine rarefies, So turn the species in their altered minds: Composed by calms and discomposed by winds.
Birds, cows, and lambs.	From hence proceeds the birds' harmonious voice; From hence the cows exult, and frisking lambs rejoice.

VIRGIL'S "GEORGICS," DRYDEN'S TRANSLATION.

Various signs of rain.	A boding silence reigns Dread through the dim expanse; save the dull sound That from the mountain, previous to the storm, Rolls o'er the muttering earth, disturbs the flood, And shakes the forest leaf without a breath. Prone to the lowest vale aerial tribes Descend; the tempest-loving raven scarce Dares wing the dubious dusk; in rueful gaze The cattle stand, and on the scowling heavens Cast a deploring eye; by man forsook, Who to the crowded cottage hies him fast, Or seeks the shelter of the downward cave.

THOMSON.

APPENDIX.

Bibliography of Weather Lore.

Abercromby, Hon. Ralph, and W. Marriott. *Popular Weather Prognostics.* London. 8vo. 43 pp.

Adams, George. *A Short Dissertation on the Barometer, etc.* London. 8vo. 1790. 60 pp.

A Descant upon Weather Wisdom. Anon. London. 8vo. 1845. 32 pp.

Allan, Wilfrid. *Weather Wisdom from January to December.* London. 12mo. 71 pp.

Angling, A Concise Treatise on the Art of, to which is added Prognostics of the Weather Independent of the Barometer. London. 8vo. 1810. 200 pp.

"Animal Weather Lore in America." Article in *Knowledge.* April 1886.

Aratus. See Prince, C. L.

Aratus, The Skies and Weather Forecasts of. Translated by E. Poste, M.A. London. 8vo. 1880.

Aratus: The Phenomena and Diosemeia. Translated by John Lamb, D.D. London. 1848.

Aristotle's Works: *Meteorology.* Translated by T. Taylor. London. 9 vols. 4to. 1812.

Bacon, Francis Lord. Works collected and edited by J. Spedding and others. 8 vols. 8vo. 1857-72.

Barometer, An Account of the, with Rules for Judging of the Changes of the Weather. London. 12mo. N.D. 11 pp.

Bohn, H. G. *A Polyglot of Foreign Proverbs.* London. 8vo. 1857.

Bohn, H. G. *Handbook of Proverbs, Comprising Ray's Collection.* London. 1855.

Brand's Popular Antiquities. Bohn's Edition. 3 vols. 1853.

Bucelini. *Historiæ Universalis auctarium.* Augustæ. 1658.

Buchleri, Joann. *Gnomologia.* Coloniæ. 1662.

Burke, V. R. *Sancho Panza's Proverbs.* 8vo. 1872.

Caballero, Fernan. *Cuentos, oraciones, adivinas y refranes.* Madrid. 12mo. 1877.

Capron, J. R. *The Rainband.* London. 8vo. 1886. 30 pp.
Chambers, G. F. *Weather Facts and Predictions.* 8vo. 1868. 16 pp.—Another Edition. 1877. 44 pp.
Chambers, Robert. *Book of Days.* 2 vols.
Chambers, Robert. *Popular Rhymes of Scotland.* London. 8vo. 1847.
Claridge, John (shepherd). *Rules to Judge of the Changes of the Weather, Grounded on Forty Years' Experience (Chiefly a Commentary on the Shepherd of Banbury's Rules).* 12mo. 1764.
Clouston, Rev. Charles. *An Explanation of the Popular Weather Prognostics of Scotland.* Edinburgh. 8vo. 1867. 53 pp.
Collins, John. *Dictionary of Spanish Proverbs.* 1823.
Companion to the Weather Glass. Edinburgh. 12mo. 1796. 118 pp.
Coremans. *L'année de l'ancienne Belgique.* Bruxelles. 1843.
Criswick, H. T. C. *The Agriculturist's Weather Guide.* London. 8vo. 1863.

Dallet, G. *La prévision du Temps.* Paris. 8vo. 1887. 336 pp.
Denham, M. A. *A Collection of Proverbs and Popular Sayings, Relating to the Seasons, the Weather, etc.* Printed for the Percy Society. 1846.
Denham, M. A. *Manners, Customs, Weather Proverbs, etc., of the North of England.* 1851.
Dickson, H. N. "Weather Folk-Lore of Scottish Fishermen." Article in *Journal of Scottish Meteorological Society.* 3rd Series. No. 6. 1888.
Dickson, H. N. "Weather Folk-Lore." Article in *Journal of Scottish Meteorological Society.* Vol. viii., p. 349.
Digges, L. *Prognosticacion Everlausting of Ryght Goode Effecte.* London. 1596. B.L.
Drôme, Mathieu de la. *Prédiction du Temps.* Paris. 8vo. 1862. 58 pp.
Dudgeon, P. "List of Proverbs," *Jour. Scot. Met. Soc.* May 1893.
Dunwoody, H. H. C. *Weather Proverbs.* United States War Department. Washington. 8vo. 1883.
Dyer, T. F. Thistleton, M.A. *English Folk-Lore.* London. 8vo. 1884.

Empson, C. W. "List of Weather Proverbs," *Folk-Lore Record.* Vol. iv., p. 127.

Folk-Lore Journal and *Folk-Lore Record.* London.
Fonvielle, W. de. *La Prévision du Temps.* Paris. 8vo. 1878. 102 pp.
Forster, Thomas. *Pocket Encyclopædia of Natural Phenomena.* London. 8vo. 440 pp.
Forster, Thomas. *Researches about Atmospheric Phenomena.* London. 8vo. 1823. 442 pp.
Forster, T., M.B. *The Perennial Calendar.* London. 8vo. 1824. 804 pp.
Fryer, J. A. *Weatherwise.* Bristol. 12mo. 1846. 116 pp.
Fuller, Thomas, M.D. *Gnomologia: Adagies and Proverbs.* 12mo. 1732.

Gutch, J. W. G. *Quarterly Journal of Meteorology.* London. 8vo. 1842-3.

Hampson, R. T. *Medii Ævi Kalendarium.* London. 2 vols. 1841.
Heap, G. *The Weather and Climatic Changes.* By "Observator." London. 12mo. 1879. 75 pp.

Henderson, J. *Meteorography.* 8vo. 1841. 26 pp. 46 plates.
Hesiod. *Works and Days in English Verse.* Chapman.
Hildebrandsson, H. H. *Samling af bemärkelsedagar, etc.* Swedish. 1883.
Hone's Works: *Every-Day Book, Table Book, and Year Book.* London. 4 vols. 1839.
Husbandman's Practice, The; or, Prognostication for Ever. London. 12mo. 1663.

Jackson, Georgina. *Shropshire Folk-Lore.* London. 1883.
Jenyns, Rev. Leonard, M.A. *Observations in Meteorology.* London. 8vo. 1848. 415 pp.
Jenyns, Rev. Leonard, M.A. *St. Swithin and other Weather Saints.* Bath. 8vo. 1871. 30 pp.

Kalendrier perpétuel aux bons laboureurs et Almanach pour l'an de grâce. Rouen. 1678.
Körte, W. *Die Sprichwörter der Deutschen.* Leipzig. 1861.

Le Roux de Lincy. *Le livre des Proverbes Français.* Paris. 1842.
Lloyd, L. *Diall of Daies.* London. 1590.
Lowe, E. J. *A Treatise on Atmospheric Phenomena.* London. 8vo. 1846. 376 pp.
Lowe, E. J. *Prognostications of the Weather.* London. 8vo. 1849.

Mann, R. J., M.D. *The Weather.* London. 8vo. 1827. 440 pp.
Marriott, W. See Abercromby.
Maudsley, Athol. *Nature's Weather Warnings.* London. 8vo. 1891.
Merle, Rev. William. *The Earliest Known Journal of the Weather: 1337 to 1344.* London. 4to. 1891.
[Latin MS. in the Bodleian Library, Oxford. Translated by Miss Parker. Published by G. J. Symons.]
Merryweather, Dr. *An Essay Explanatory of the Tempest Prognosticator.* [An instrument worked by leeches.] London. 1851.
Meteorological Reports. From the Natural History Transactions in Northumberland and Durham. 1864 to 1872.
Mills, John. *An Essay on the Weather.* London. 8vo. 1773. 127 pp.
Mitchell, Sir. A. "On the Popular Weather Prognostics of Scotland," *Edin. New Phil. Jour.* XII. 1860.
Murphy, Patrick. *Meteorology.* London. 8vo. 1836. 277 pp.

Notes and Queries.
N. S. *A Tract Concerning the Weather.* London. 1653. 50 pp.

Oihenart, A. *Proverbes Basques.* Bordeaux. 1847.

Pasqualigo, Cristoforo. *Raccolta di Proverbi Veneti.* Venezia. 1858.
Pearce, Alfred J. *The Weather Guide Book.* London. 8vo. 1864. 141 pp.

Philotechnus. *Remarks on Barometer Scales.* Edinburgh. 8vo. 1814. 51 pp.
Pliny. *Historie of the World.* Translated by P. Holland. London. 1634.
Pluquet, F. *Contes populaires, etc., de l'arrondissement de Bayeux.* Rouen. 1834.
Pocket Companion to the Board of Trade Storm-Warning Signals. 12mo. 14 pp.
Poëy, André. *Comment on observe les nuages, etc.* Paris. 8vo. 1879.
Pointer, M. A. *A Rational Account of the Weather.* 8vo. 1723. 224 pp.
Prévision du Temps. Paris. Annuaire. 1887. 12mo.
Prince, C. Leeson. *Observations upon the Climate of Uckfield, Containing a Translation of "Aratus."* London. 8vo. 1871. 341 pp.
Proverbes (2228). Brussels. 12mo. 1854.
Proverbes et dictons agricoles de France. Paris. 1872.
Proverbial Folk-Lore. Dorking. N.D.

Ray, J. *Proverbs.* See Bohn, H. J.
Reinsberg-Duringsfeld, Baron Von. *Das Wetter im Sprichwort.* Leipzig. 1864.
Remarks on Revolving Storms. London. 8vo. 1851. 27 pp.
Roper, A. W. *Weather Sayings, Proverbs, and Prognostics, Chiefly from North Lancashire.* 8vo. 1883. 34 pp.
Rosa, G. *Dialetti, costumi e tradizioni delle provincie di Bergamo e de Brescia.* Bergamo. 1857.
Russel, Hon. F. A. R. "On Cirrus and Cirro-Cumulus." Article in *Quarterly Journal of Meteorological Society.* Vol. ix., No. 47.

Samarani, Bonifacio, Professor. *Proverbi Lombardi.* Milano. 1870.
Saul, Edward. *An Historical and Philosophical Account of the Barometer.* London. 8vo. 1766. 107 pp.
Sawyer, Frederick E. *Sussex Folk-Lore and Customs Connected with the Seasons.* Lewes. 8vo. 24 pp.
Sawyer, Frederick E. *Sussex Natural History Folk-Lore and Superstitions.* Brighton. 8vo. 1883. 16 pp.
Saxby, S. M., R.N. *Saxby's Weather System.* London. 8vo. 1864.
Shepherd of Banbury. See Claridge, John, and Taylor, Benjamin.
Shepherd's Kalendar; or, Countryman's Companion. London. N.D.
Smith, Rev. A. C., M.A. *On Wiltshire Weather Proverbs and Weather Fallacies.* 8vo. 1873. 29 pp.
Steinmetz, Andrew. *Everybody's Weather Guide.* 8vo. 1867. 24 pp.
Steinmetz, Andrew. *Sunshine and Showers.* London. 8vo. 1866. 432 pp.
Steinmetz, Andrew. *Weather Casts and Storm Prognostics.* London. 8vo. 1866. 208 pp.
St. Felix, Marquis de. *Meteorologie du Cultivateur.* Toulouse. 8vo. 1870. 152 pp.

Strachan, R. *Principles of Weather Forecasts.* London. 8vo. 1868. 24 pp.

Swainson, Rev. C., M.A. *A Handbook of Weather Folk-Lore.* London. 8vo. 1873. 275 pp.

Swainson, Rev. C., M.A. *Folk-Lore of British Birds.* 8vo. 1885.

Taylor, Benjamin. "Weather Wisdom." Article in *Victorian Magazine.* December 1891.

Taylor, Joseph. *The Complete Weather Guide, including the Shepherd of Banbury's Rules.* London. 12mo. 1814. 160 pp.

"The Clerk Himself." *Weather Warnings for Watchers.* London. 8vo. 1877. 96 pp.

Theophrastus. *De Ventis and De Signis pluviarum, etc.* Translated by G. Schneider. (Vol. ii. of *Theophrasti Opera.*) Leipzig. 1818.

Toaldo, T. *Essai Météorologique.* Translated by M. Joseph D'Aquin. Chambery. 8vo. 1784.

Toplis, John. *Observations on the Weather.* London. 8vo. 1849. 108 pp.

Tusser, Thomas. *Five Hundred Points in Good Husbandry.* 4to. 1812.

Virgil. *Bucolicks and Georgicks.* Translated by Martyn. 2 vols. 8vo. 1749-53.

Weather Book, The: Three Hundred Plain Rules for Telling the Weather. London. 12mo. 1841. 64 pp.

Weather Wisdom. London. 12mo. 1860. 16 pp.

Whistlecraft, Orlando. *Rural Gleanings.* London. 12mo. 1851. 270 pp.

Willsford, Thomas. *Nature's Secrets.* London. 1665.

Wing's Ephemeris for Thirty Years. London. 12mo. 1569.

Woollams, Elizabeth. *What do the Leeches say?* Cir. 1859. 12 pp.

Wormu, Olai. *Fasti Danici.* Hafniæ. 1643.

Wurzebach, C. *Die Sprichwörter der Polen.* Wien. 1852.

INDEX.

REST PRAY SLEEP